Fragile Land

**Books are to be returned on or before
the last date below**

Fragile Land

Scotland's Environment

Auslan Cramb

POLYGON
Edinburgh

For Chae and Juliet

© Auslan Cramb, 1998

Polygon
22 George Square, Edinburgh

Typeset in Baskerville
by Norman Tilley Graphics, Northampton, and
printed and bound in Great Britain by
The Cromwell Press, Trowbridge, Wilts

A CIP record for this book is available from the British Library

ISBN 0 7486 6228 6

The right of Auslan Cramb
to be identified as author of this work
has been asserted in accordance with
the Copyright, Designs and Patents Act 1988.

The Publisher acknowledges subsidy from

THE SCOTTISH ARTS COUNCIL

towards the publication of this volume.

Contents

·

Foreword

·

Magnus Magnusson, KBE

·

A S WE APPROACH THE new millennium, all of us who care
about Scotland's precious environment face important
decisions and major challenges. If we are to safeguard and
enhance the wildlife and landscapes, the special places and
the 'wider countryside' we have celebrated for centuries in
poetry and song, we shall have to take greater responsibility
for our actions, both individually and collectively.

Auslan Cramb is a distinguished journalist, an award-
winning writer on environmental issues. He has always had
a special interest in, and feeling for, Scotland's natural
heritage. He and I have had frequent occasion to discuss the
environmental challenges on which it has been his business
to comment. As chairman of Scottish Natural Heritage
(SNH), the government's environmental agency in Scotland, I did not always agree with his trenchant views – but I
always respected them, and often learned from them.

Fragile Land portrays, for lay and learned alike, Auslan's
deeply-felt views about the state of the Scottish environment.
He examines the health of Scotland's towns and cities as well
as Scotland's mountains, waters and wildlife. He presents a
penetrating analysis of the serious problems in our uplands,
of the inefficient use of energy in our homes and businesses,
and of the dangerous lack of knowledge about our marine
environment – and who can gainsay him?

He also points to the need for better and more widespread environmental education, to ensure that all our people will have the opportunity to understand the essentially fragile nature of the land in which we live: only through greater understanding can we hope to ensure the long-term protection of the environment which sustains us all.

Auslan's conclusion – which I share wholeheartedly – is that concerted action on the environment is now required as never before: concerted action by everyone in Scotland, old and young alike, and not just by the national and local authorities and agencies of the voluntary conservation agencies.

Fragile Land is an important and timely book, as well as being an excellent read. The new Scottish Parliament will bring more decision-making north of the Border, and this will have an immense significance for our environment. It is time for a fresh look at problems and issue of today, and a fresh determination to find new solutions for the challenge of tomorrow; I hope this excellent book will be read by all those who will determine the way in which the Scottish environment of the twenty-first century is treated.

.

Acknowledgements

.

THIS BOOK WOULD NOT have been possible without the work of the numerous scientists, naturalists and conservationists who have been studying the Scottish environment for many years.

The 'resource' sections in Chapters 2–5 and 7 are based largely on the Scottish Natural Heritage report, *The Natural Heritage of Scotland: An Overview* (1995). They also draw on the *State of the Environment 1996* report from the Scottish Environmental Protection Agency, and the *State of the Scottish Environment 1991* report commissioned by Scottish Wildlife and Countryside Link from T. C. D. Dargie and D. J. Briggs. Numerous papers and reports from Friends of the Earth have been used, along with studies by the Royal Society for the Protection of Birds and the Worldwide Fund for Nature.

Too many people to mention have helped, but I would like to name some of those who have assisted not only with this book but also with my understanding of the Scottish environment over the years. They are: Dick Balharry, Wattie Barbour, David Begg, Tom Brady, Richard Cooke, Roger Crofts, Dave Dick, Roy Dennis, Kevin Dunion, John Goodlad, Patrick Gordon Duff Pennington, Ron Greer, Stuart Housden, Jim Hunter, Sir John Lister Kaye, Nicholas Luard, Magnus Magnusson, Martin Mathers, Prof. Alan McKinnon, David Minns, Dave Morris, Simon Pepper, Jonathan Porritt, Derek Pretswell, Bill Ritchie, Michael

Scott, Chris Smout, Des Thompson, Paul van Vlissingen, Dr Adam Watson, Alan Watson, Drennan Watson and, last but not least, David Westwood. A special thank-you also to award-winning photographer Laurie Campbell whose excellent pictures illustrate the book.

Finally, I have to thank my wife, Catriona, for looking after our children on what must have seemed like countless evenings and weekends when I was trying to 'do my book'.

Abbreviations

ACE	Association for the Conservation of Energy
BSE	bovine spongiform encephalitis
CAP	Common Agricultural Policy
CEC	Crown Estates Commission
CFC	chlorofluorocarbon
CHP	Combined Heat and Power
CZM	Coastal Zone Management
DoE	Department of the Environment
EAS	Energy Action Scotland
EU	European Union
FoE	Friends of the Earth
HSE	Health and Safety Executive
LFA	Less Favoured Area
NGO	non-governmental organisation
NHER	National Home Energy Rating
PCB	polychlorinated biphenyl
ppb	parts per billion
ppm	parts per million
RSPB	Royal Society for the Protection of Birds
SEPA	Scottish Environment Protection Agency
SNH	Scottish National Heritage
SPA	Special Protection Area
SRO	Scottish Renewables Order
SSSI	Site of Special Scientific Interest
VOC	volatile organic compound
WHO	World Health Organisation

1
·
Introducing Scotland
·

We are apt to view with pleasure a rugged Highland landscape
and think we are here away from the works of the mind and
hand of man, that here is wild nature. But more often than not
we are looking at a man-made desert: the summits of the hills
and the inaccessible sea cliffs alone are as time and evolution
made them.

Sir Frank Fraser Darling, *Natural History*
in the Highlands and Islands, 1947

NOTHING MUCH HAS changed in the fifty years since the
ecologist Fraser Darling remarked that Scotland's bare,
heather hills were a man-made desert. Sadly, his claim did
not have much resonance with the government of the day,
far less with the public. Today, on the edge of a new millennium, Scotland is still, as Shakespeare had it, 'almost afraid
to know itself'.

In a sense, we are all guilty. There remains, at home and abroad, a public perception of Scotland as overwhelmingly clean and green. Its reputation, whether for tourism or natural produce – beef, salmon and whisky – is founded on an image of natural beauty, pure waters and unsullied pastures. We trade, in every way, on our mountains, rivers, lochs and shores, and we are likely to conclude that Scotland is looking good for the next thousand years.

Asked by a foreigner why we love Scotland, we say it is because of the Highland mountains and the rolling Borders hills, the small pub on the sea loch, the tang of the seaweed, the prawns, the lobsters, the water of life.

My own ideal of Caledonia stern and wild is scattered through the Highlands and the islands. I can find it in the mountains of Glencoe, on a trout loch on Barra, on the beaches of Coll and the rugged west coast of Mull. I can find it also on the shores of Loch Awe, on the islands of Loch Maree, in the heart of a Cairngorm pine forest, and among the voes and skerries of sea-sprayed Shetland.

But in each of these places, there is damage and degradation below the surface beauty. To a large extent, we have habituated the Scotland we have; the changes on the landscape have been so slow, taking more than a single lifetime, that we no longer find anything odd about a treeless Glencoe, about a Colonsay dominated by holiday homes, about the unsightly hill tracks in the Cairngorms, the soil erosion caused by too many sheep on Shetland, and the grazing damage caused by deer and goats in our ancient oakwoods.

In the urban environment, the damage is often insidious. Here, as in every other developed country, the rain washes oil and heavy metals off roads and out of building yards into rivers and streams. Acid rain makes our lochs fishless; global warming threatens our alpine plants; air pollution kills hundreds of people every year; and sewage is dumped untreated into the marine environment.

It is only when we are confronted with them in the most direct fashion that we are apt to recognise some of our myriad problems. We are, for example, concerned about traffic congestion on our roads, about damp, inefficient housing, and about stagnating and deprived housing schemes. In clean, green Scotland.

Scotland is a fragile land, if only we knew it, and the purpose of this book is to look beneath the postcard image that informs so much sentimental thinking about the country. We should not enter the next millennium dragging with us dangerously simplistic notions formed by the observation that 'it must be healthy because it still looks attractive'.

Some species are under-protected, others are over-protected. Some require recovery strategies, others – like the red deer and perhaps even the grey seal – reduction strategies. Some new roads are justified, others are an environmental outrage.

Striking the right balance – a task for government, agencies, planners and individuals – is problematic. We must overcome the familiar, but out-dated, mode of thinking which sets 'environment' against 'development'. The goal can be particularly elusive when the urban public – 80 per cent of the Scottish population – is distanced, physically and intellectually, from the natural world. Unfortunately, many people living in cities think that the 'environment' begins outside the urban area, and see no role for themselves in tackling global warming, or transport congestion. They may also have been unduly influenced by environment groups which depend on high-profile actions and campaigns for support, membership and survival, and therefore espouse a simplistic 'save the whale' mentality to all areas of the natural world.

Would the urban majority, for example, condone the shooting of thousands of geese? Would they eat them? There are now tens of thousands of geese over-wintering in Scottish farm fields. They are protected in their Arctic breeding grounds, as well as their winter feeding areas on Islay and the Solway Firth where the farmland has been improved through agricultural subsidies. So, the birds fly north in springtime – in the best of fettle – leaving Scottish Natural Heritage (SNH) to compensate farmers and crofters for the damage they have caused – the torn grass, the droppings and the fields paddled into mud. In 1995, the payments on Islay alone amounted to £300 000, or £9 a bird. Is it, perhaps, time to allow a cull, and to allow goose meat to be sold on the open market? Our objective in such cases should be population viability and species diversity, not sheer numbers.

We may also have problems looming over access to the land, and, in particular, to Scotland's 284 Munros (peaks over 3000ft). People have more leisure time than ever, and the increasing use of mountain areas means that more than 300 mountain rescues are now being carried out each year, compared to less than fifty in the 1960s. Thirty years ago, the Cairngorms were four hours from central Scotland; today they are just two-and-a-half hours away. Torridon is four hours from the populous central belt while once it was a drive of more than seven long hours.

Most day-trippers and holidaymakers visiting Scotland say they are attracted by the scenery. They visit the countryside to celebrate, and get their slice of, the natural environment, to climb, ski, walk and fish. But will the day come when there are too many people in the most fragile areas? And will we have to limit access, or even charge people to enter wild landscapes?

Elsewhere, in the arena of business and industry, is it time to introduce more radical market tools which reward energy efficiency and encourage the reuse and recycling of resources? Should we, for example, reward business for cutting its use of electricity, instead of granting it low tariffs for gluttony? And should we switch the burden of taxation away from the individual and on to resources, so that the true environmental costs of, for example, fossil fuels and other finite natural resources are realised? And could we, at the same time, make business more competitive and efficient?

Should we work to improve the living environment of the 100 000 people in Glasgow's four peripheral housing estates, and those in Wester Hailes in Edinburgh, Ferguslie Park in Paisley and Whitfield in Dundee? In these localities local initiatives are tackling unemployment, training, housing, crime and health, but are fuelled more on energy and optimism than real resources. I hope the reasons for asking such questions will emerge in later chapters. At this point, three examples of Scotland's 'unseen' problems may help to illustrate the challenge ahead.

The uplands – The most obvious and pressing problem confronting that precious hinterland of mountains, lochs and rivers is bad management in the form of over-grazing by sheep, deer and, sometimes, rabbits. The high deer numbers are supported by privately owned sporting estates whose value is calculated by the number of stags shot during the stalking season. The high sheep numbers are supported by EU subsidies, which pay farmers on a 'headage' basis – the more sheep you have, the more money you make, the worse the habitat damage.

Grazing pressure, in combination with poorly executed heather burning (muirburn) for the benefit of grouse or sheep, has created Fraser Darling's 'deserts' on either side of the A9. The same is true between Perth and Inverness, on either side of the A82, from Tyndrum to Glencoe, on either side of the A708, from Selkirk to Moffat, and in many other places.

For several centuries, over-grazing has stopped trees and other vegetation from growing, and has created our typically smooth moors and bare mountains, only occasionally whiskered by a single rowan or birch clinging to areas which are inaccessible to the grazing mouths. John Lister Kaye, the naturalist, called this landscape 'Mamba' – miles and miles of bugger all.

The loss of woodland means the loss of mammals, birds, insects, fruit, a decline in soil fertility and, potentially, serious soil erosion. The real wealth of our uplands should be the soil–vegetation complex, not the deer and sheep which extract goodness from the soil every year, and put little back.

At a best guess, Scotland has around 350 000 red deer. Historically, they were woodland creatures, but have evolved to live on the open moor. In the process they have become the world's smallest, and lightest, red deer.

The effect of high deer numbers can be seen in many places, notably in the Cairngorms. On Mar Lodge estate – now, thankfully, owned by the National Trust for Scotland and not by a private individual – there is a dwarf forest of many millions of pine trees hidden below the heather. Some are only five inches tall but fifty years old. They have been kept at that height by the herds of red deer so admired by the passing walkers. If the trust succeeds in reducing deer numbers to a sustainable level, the forest will flourish again. With the trees will come enhanced biodiversity, in the form of the pine marten, crested tit, capercaillie, crossbill, red squirrel, wood ant, pinewood hoverfly, bird cherry, rowan, birch, aspen and others. Perhaps even the tourist, naturalist and scientist.

Energy – Scotland has always been rich in sources of energy, ranging from peat, coal, gas and oil, to wind, waves and the sun itself. We currently produce a surplus of electricity north of the Border, with nearly 40 per cent coming from our large, coastal nuclear power stations. This source of energy does not cause serious pollution on a daily basis, but produces a legacy of highly radioactive waste that will be with us for thousands of years.

Coal was once the main source of power in Scotland, with demand peaking at 43 million tonnes in 1913. Today, we produce less than 10 million tonnes, and most of that comes from unsightly open cast mines. But the coal-fired power stations – Longannet is the biggest – remain a major source of pollution, and contribute to acid rain and global warming.

The other main fossil fuels, the oil and gas which were discovered in the North Sea in the 1960s, are being used at a rate which, in the near future, we are likely to find remarkable. Every day we burn in seconds the oil which was created over millions of years.

In recent years, there has been a shift, driven by economic considerations, towards more efficient, and less-polluting gas-fired power plants. This has delivered some environmental benefit by reducing emissions of the carbon dioxide (CO_2) which is responsible for man-made global warming.

Yet none of our main sources of power can be described as sustainable in the long term, and most of our power plants remain hugely inefficient.

Our power generation could be much 'cleaner', but our slovenly ways have been encouraged by our wealth in mineral deposits. Sweden, by contrast, has been encouraged by its lack of coal and oil to develop a much more efficient power generation regime. While our power stations waste great quantities of heat to the air, Sweden has for decades used the heat generated in power plants to heat homes and office buildings.

Our utilities also produce a significant surplus of electricity, with major exports to England, and have seen little advantage in turning to renewable sources of energy. This is unfortunate because Scotland has the best potential in Europe for wind and wave power projects.

Our profligacy with energy extends to our homes. While Scandinavian building standards produce warm, comfortable homes in a harsh climate, our housing is among the most inefficient in the developed world. It has been estimated that domestic energy use in Scottish homes could be cut by 50 per cent with very simple technologies, including better insulation. It should also be a matter of national concern that there are 750 000 households in Scotland which are inadequately heated in winter because their occupants cannot afford the unreasonable electricity bills. Fuel poverty is one of Scotland's most pressing urban problems.

The marine environment – The sea has an integral role to play in forming positive attitudes towards Scotland. But if we are ignorant of some terrestrial issues, we know next to nothing about the management of the more complex marine environment.

We know very little, for example, of the habits and habitats of the nine species of whale which may breed in Scottish waters; we don't know where the sea trout goes on its annual migration, and we don't know why it is disappearing from our waters. We don't know, for certain, what effect commercial fishing is having on important stocks, although we allow the seabed to be dragged several times a year by heavy fishing gear.

Our shores are home to 45 per cent of the world's grey seals and 38 per cent of Europe's common seals, and our salmon anglers and farmers and inshore fishermen are convinced the cute mammals are out of balance with their environment. But we do not have the scientific knowledge – because we do not fund the research – to tell us if there should be a seal cull. And if we decided there should be a cull, it is likely it would be prevented by an emotional, and irrational, public outcry.

Meanwhile, we continue to use the marine environment as a convenient waste repository. There are around 1000 sewage discharges in Scotland, and industry and traffic release toxic pollution to the North Sea and the Atlantic. We have been dumping in the sea for many decades, a fact illustrated in 1995 when thousands of incendiary devices from bombs dumped after the Second World War were washed on to the beaches of the Clyde coast.

And we are only now, after centuries of exploiting the sea, beginning to think of mechanisms and partnerships to look at the strategic management of a serrated coastline of 11 800km which contains some of the finest wildlife habitats in Europe. However, Scotland has no marine nature reserves, and fewer marine research institutes than it did in the 1980s.

Each of the examples above can be considered in terms of 'sustainability', the buzz-word of modern environmental thinking which directed discussions at the Earth Summit in Rio in 1992, and which will resonate throughout this book. But on the edge of the next millennium, it is difficult to see how the concept of sustainability has been enshrined in the policies of local and national government.

So far, my potted description of this fragile land has been pessimistic in tone. But the story of the Scottish environment need not be depressing! Scotland, if it made the right choices, could be on the cusp of sustainable development. The degradation of our cities, forests, hills, rivers and lochs is not so far advanced that it cannot be reversed.

The man-made deserts may have spread a little further

since Fraser Darling's day, but they can be made to bloom. The unplanned, road-dominated transport system can be made to work; strategic planning can alter our energy production regime and energy efficiency in the home and the workplace can be improved many times over. Our cities have not spawned the street children of Bogota, our traffic does not generate the choking fumes of Mexico City, and our intensive agriculture is not on the scale of the Russian models which have destroyed or depleted huge areas of fresh water.

And yet, because we are not, as individuals, particularly close to nature, our condition is best summed up as a mixture of crisis and opportunity. The aim of the final section of this chapter is to explain our current condition by sketching many thousands of years of Scottish history!

Before the First Settlers

Some 30 000 years ago, Scotland had a periglacial landscape in which woolly mammoth, woolly rhinoceros and reindeer grazed on open parkland. There then followed 10 000 years of cooling which culminated with the last Ice Age, and an ice cap up to 2000m thick.

The ice lasted for 5000 years, and when the climate at last began to warm the retreat of the glaciers, which had carved new valleys and smoothed the landscape, was slow and irregular. The disappearing ice left a messy landscape, but one which would eventually be re-colonised by plants and animals. The evidence of glaciation is all around us: the melting ice dropped great boulders – known as erratics – on ridges and hills, trailed eskers (ridges of sand and gravel) in its wake; and exposed the drumlins (smooth teardrop-shaped landforms) which it had created as the ice expanded.

Around 13 500 years ago, with the ice still retreating, the new vegetation would have been Arctic in character, varying, according to the rainfall, from blaeberry and willow in the west, to steppe in the east. The open landscape would have been home to elk, reindeer, giant fallow deer and arctic lemming.

9

As the climate warmed further, shrub cover developed and the first birch forests arrived by 8000 BC. Eventually, the Arctic communities became restricted to the high hills and great forests of birch, oak, elm, hazel and pine took over. Trees spread to cover nearly 80 per cent of Scotland, although the appearance of each new species took time. Birch, for example, appeared 3000 years before ash.

One of the most crucial developments, as the ice retreated, was the rise in sea levels which created the English Channel. What was to become Britain, then became an island, and its flora and fauna were isolated from the continent for the first time.

The Arrival of People: Forest Clearance

The first human imprint was left by Mesolithic hunter-gatherers, around 7000 BC, on the Hebridean island of Rum. But it was not until 4000 BC that Neolithic farmers began clearing birch woodland and tilling the light upland soils with their stone tools. Woodland cover was probably at its maximum extent 5000 years ago, and has been in decline ever since, largely due to climate change. Man's influence on the great forests accelerated through the Bronze Age, and the loss of the habitat led, eventually, to the extinction of the lynx, elk, wild horse, wild boar, beaver and brown bear.

The greatest single period of forest clearance – and perhaps the biggest change in the history of the Scottish landscape – came during the Iron Age, roughly 2500 years ago, around the time of the arrival of the Celtic people from Ireland. By the time the Romans appeared in AD 43, the lowlands were already largely devoid of forest and grazing animals were being introduced in much greater numbers than before. Fire, which had been used since the mid-Bronze Age to clear woods for pasture, was the tool employed by people hungry for land. Clearing the forest also served to keep the wolves at bay.

Following the Roman withdrawal in AD 406, there was a

brief period of forest expansion in lowland Scotland and in relatively populous areas such as Fife. But the recovery lasted no more than a few centuries, and rapid population growth from the Middle Ages onwards put paid to the expanding tree cover and opportunities for wildlife. By the fourteenth century, there were two million sheep in the lowlands and wood was scarce around towns. At the same time, several centuries of animal grazing in the Highlands, coupled with the deliberate destruction of forest and the growth of peat, had dramatically reduced forest cover in the north.

Throughout the eighteenth century and during the early nineteenth century, the population of the Highlands was high and the people kept large numbers of goats. Historical documents indicate that overgrazing was already restricting forest growth in many areas. The periods of forest conservation have been few and far between, although some oak woods on the west coast were managed responsibly in the seventeenth and eighteenth centuries to provide the charcoal needed to smelt iron, and the bark needed for tanning leather. However, many ironmasters simply cleared forests and left the saplings to be grazed by animals.

In the nineteenth century, when coal replaced charcoal and artificial chemicals replaced the tannin of oak bark, even the woods which were being coppiced to maintain their growth went into decline. On balance, historians have concluded that the use of timber and tan bark during the industrial revolution probably did more good than harm. But the increased importation of cheap foreign timber, which replaced Scottish birch bobbins and pine planks, reduced the commercial value of forests and caused further decline.

The Day of the Sheep
·

Sheep have been one of the major shapers of Scotland and continue to be so. The first large flocks were run by medieval abbeys, and the conversion of deciduous woodland into sheep grazing began as long ago as the twelfth century. In the seventeenth century, 100 000 goatskins were exported from Inverness to London in one year, and the damage that the animals caused to woodland was well documented. From the mid-eighteenth century, great swathes of the Highlands were converted by their owners into cattle ranches, and then into sheep runs – leading towards the Clearances and the dominance of the ewe on open moors. Decades of heavy grazing, aided by deer, helped to convert heather moorland into impoverished, coarse grass sward.

Industry, Agriculture and a Booming Population

The natural environment then, has been changing for thousands of years. But the pace has quickened exponentially in the past three centuries. In the eighteenth and early nineteenth centuries, land improvement was widespread, with crop rotations – involving turnip – being introduced for the first time. Large areas of wetland were drained in eighteenth-century improvement schemes, and in the Highlands much arable land was abandoned during the late eighteenth and early nineteenth centuries as people were moved away to make space for sheep.

When sheep farming declined in the 1870s, in the face of competition from Australia and New Zealand, many estates were converted to deer forest.

Scattered industry began with textiles and the development of the 'easy' coalfields of central Scotland. By 1759, pig iron was in production, and steam was driving the cotton mills of the Clyde in 1780. Steel production began on Clydeside in 1880. The industrial revolution brought technology, progress and pollution. The Scottish countryside had been transformed by sheep, and the Scots were transformed by the blast furnace, the steam engine and the great shipyards, coal mines and engineering plants. The population rose steeply in the countryside and the towns.

At the Act of Union in 1707 there were around one million Scots. By the time of Queen Victoria's first visit in 1842, five years after she ascended the throne, there were 2.6 million. Indeed, the Scottish population was to grow from 1.6 million in 1801 to 5.2 million in 1961. The great migration to the cities and towns had begun in earnest in the mid-nineteenth century, and, as early as 1880, 70 per cent of the population was urban dwelling. Glasgow, according to the historian T. C. Smout, was a 'squalid industrial megalopolis of textiles and engineering' during Victoria's visit. Its population of around 275 000 had multiplied twelve times since 1775. Edinburgh, meanwhile, had 138 000 residents, and its Old Town slums were severely crowded. Numbers were also growing quickly in industrial Lanarkshire, where there was said to be one pub for every twenty men.

It is probable that Victoria, like many people today, thought of Scotland as two countries: a place of romantic Highlands glens and squalid industrial heartlands. With the industry, came the pollution from furnaces, shipyards and belching chimneys which had killed all fish life in parts of the upper Clyde by 1854.

The march of progress also changed agriculture, which, as late as 1830, had been organic and highly productive. Boggy areas and wetlands had been drained, hedges of hawthorn and windbreaks of beech and sycamore had been planted throughout the countryside, and nitrogen-fixing crops like clover and beans were rotated with food crops to improve the soil. These common-sense practices began to change from the 1840s, with the importation of Peruvian guano and nitrates from Chile.

Fishing was also developing apace. Markets in the Baltic and Germany fuelled the exploitation of herring in Scottish waters between 1790 and 1914. Sailing boats gave way to steam drifters which eventually began to trawl their nets. The next phase was the development of the purse seine net and, after 1950, the sonar. These technological advances precipitated the herring stock collapse and the closure of the fishery from 1976 to 1982.

Whaling thrived in the mid-eighteenth century and survived for around 100 years, with bases at Peterhead and Dundee. However, competition from other nations led to a decline in catches, until only 100 whales were taken over a period of nine years in the mid-nineteenth century. The bowhead whale, the main catch, has not yet recovered. But the indigenous whaling industry survived in the Outer Hebrides and Shetland until the 1950s; 6000 fin, 400 blue, 95 sperm and 70 humpback whales were taken in Scottish waters between 1903 and 1929.

The Twentieth Century
·

The twentieth century has been a period of extraordinary technological advance. Agriculture has witnessed increasing mechanisation and intensification, with a consequent reduction in the rural workforce. As much land as possible has been utilised for crops, and the use of pesticides, insecticides and fertilisers has increased yields and reduced biodiversity. Hay meadows and small mixed farms have disappeared, wet areas have been reclaimed and industrial farm buildings have sprung up across the countryside.

Great areas of timber have been felled and the shortage of timber in the First World War led to the setting up of the Forestry Commission in 1919. Much of the felling during the two world wars took place in long-established woodlands, with native woodland remnants being cleared or under-planted with non-native exotic species, principally sitka spruce and larch. Large areas of hill grazing and moorland were also transformed by conifer planting, often in unattractive swathes of single-age, single-species trees.

After 1945, land-use change was effected by technical, market and economic factors, including the drop in the price of wool. Hill farming then became based on lamb production, with the stock being moved to low ground for fattening. Where sheep farming retreated, and new forests were not created, the red deer filled the vacuum.

Lowland farming has undergone a revolution in the past fifty years with production being concentrated on cereals, and then, more recently, on oilseed rape.

The marine environment has also changed this century. The space-age technology of the modern trawler has made it possible for commercial fishing enterprises to destroy stocks of haddock and cod. The seas around our coast have also been under assault from pollution carried by, among others, the Rhine, the Thames and the Forth.

Not surprisingly, perhaps, recent decades have also witnessed the establishment of government environment agencies, and the growth of a powerful voluntary environmental movement which is now led by wealthy charities such as the Royal Society for the Protection of Birds and the Worldwide Fund for Nature.

Today

Markets dictate much of what happens in the world. Thus, over-production in Europe has resulted in less pressure to expand agricultural production and commercial forestry in Britain. Arable land which once was cleared to make way for crops, is now being 'set aside' to take land out of production. Ironically, we are paying farmers to put back the landscape features which once they were paid to remove – the hedgerows and the boggy areas which are so important to wildlife. However, livestock payments are still made on a headage basis, and the management of sporting estates has changed little in a century. On a more positive note, the Forestry Commission has undergone a sea change in attitude, and environmental considerations are now at the heart of its planting schemes.

Perhaps the most obvious environmental improvements have been won in our major industrial rivers in Scotland, with salmon returning to the lower Clyde and the River Kelvin. Elsewhere, some species hunted to extinction this century, for example the sea eagle and the red kite, have been reintroduced and are breeding again in the wild.

There are signs of a conservation ethic developing in Scotland, but there is plenty to bemoan at the turn of the millennium. Most of the fertile soils are used for agriculture alone, or have been built upon, so our wildlife is restricted to relatively poor land. Soil has become degraded in some areas, native and semi-natural woodland is in poor condition and declining in many parts of the country, and important mudflat, inter-tidal habitats have been lost through land reclamation. Road improvements and house building are eating into the countryside, and the process of development has to be supported by aggregates quarried from the remotest, and wildest, areas of Scotland. Acid deposition and global warming continue apace, and our transport system is unsustainable and dominated by roads.

Britain's first White Paper on the environment, published in September 1990 – to lukewarm reviews – remarked that ever since the Age of Enlightenment we have had boundless faith in our intelligence and the healing powers of the natural world. Market principles, it said, had improved life but had also led people to question their quality of life and had produced demands for cleaner streets, better air and protection for the natural world.

The above potted history of the Scottish environment suggests we have come only lately to such concerns. In fact, the environment with which we are dealing today has become incredibly varied and complex in a very small space of time. If we say that man first walked erect twenty-five years ago, then it took twenty-four years to discover the circulation of blood, and the genetic code which governs life was only cracked nine days ago. And while it took nearly all of the twenty-five years for the human population to expand to 2.5 billion, it took just one more day to reach five billion.

Now, in a technological age in which the development of new products far outstrips our ability to understand them, far less their effect on the environment, we have the concomitant ability to make bad decisions on a daily basis. Just two centuries ago, technology was easily understood, and markets were local. The resident of a small Scottish town would source everything required for his or her life – clothes, food, building materials – within a matter of a few miles. A map of the individual's 'supply world' would cover the hinterland of the town. Today, every one of us in the developed half of the world requires the entire planet for our daily needs. Food, clothing, motor cars, electronic equipment and household materials are sourced from overseas. It is easy to see, in terms of energy efficiency and resource use, that such a system is bad for the world environment.

Yet the dominance of multinational companies means that decisions are taken for the world, often by a relatively small number of individuals and organisations in a relatively small number of countries. For example, the discovery of inert, cheap and non-toxic gases, called chlorofluorocarbons (CFCs), was hailed as a major breakthrough. They were used as propellants and coolants in aerosols, refrigeration and air-conditioning. The scientists who approved them had no idea that their long-lasting properties would allow them to reach the upper atmosphere, where they would react with sunlight and destroy the protective ozone layer which screens life from the cancer-causing effects of ultraviolet radiation.

Similarly, lead was introduced to petrol to make engines more efficient, and pesticides were used in vast quantities on farmland in the 1950s to increase our food output. In each case there were dire consequences for us, and for the wildlife around us. Unquestionably, we are still making similar mistakes.

Incidents such as these have, at least in theory, inspired us to adopt the 'precautionary principle'. This means that if we think our actions might be damaging, we should stop them, without demanding absolute scientific proof. The same incidents suggest that Scotland is part of a much bigger whole, and cannot be viewed in isolation. Many of this country's key environmental problems will only be solved by international co-operation in Europe and, in some cases, by global action.

The greatest cause for optimism can be found in the fact that people, more than ever, claim to be concerned about the environment, even if wealth creation and personal comfort still come first. We may not always practise what we preach, but we have signed up to the idea of enhancing our wildlife and wild places. Membership of environmental groups is rising, and the biggest organisations have more supporters than the main political parties.

In a survey conducted before Labour's General Election

victory in 1997, 55 per cent of respondents said they would trust environment groups first, 20 per cent trusted scientists, and just 3 per cent trusted government and its agencies. However, in the immediate aftermath of the Labour win there was an upwelling of optimism among environmentalists. Experts and amateurs expressed the hope that the change of government would, at least, usher in a new culture of social and environmental concern. There have been numerous positive signs since, particularly on the issue of transport. There were also some early moves on energy efficiency, with a reduction in VAT on home insulation materials.

In a Scottish context, there was a remarkable reversal of the previous government's policy on land ownership when Scottish Natural Heritage was instructed to become involved in a public sector bid for Glenfeshie estate in the Cairngorms. The bid failed when the property was purchased by a Danish millionaire, but the principle of government support for community/public ownership had been established.

Another important step was taken when Scotland's first national park was announced. It is intended to protect Loch Lomond and the Trossachs and should be in place early in the new millennium – only sixty years or so since the first demand for a parks system to protect our finest wild lands.

In the longer term, radical change will depend on the education of environmentally aware generations. One survey designed to test environmental knowledge showed that people thought they understood lifestyle issues such as transport or water quality. But only 5 per cent knew much about forestry, a similar figure thought they understood nuclear power, and less than 10 per cent knew about acid rain and global warming. Of course, ignorance is not the prerogative of the man in the street. Mrs Thatcher once famously said: 'It is exciting to have a real crisis on your hands [the Falklands War], when you have spent half your political life dealing with humdrum issues like the environment.'

A more enlightened commentator, Gro Harlem Brundtland, the former Prime Minister of Norway, said in 1987: 'If we take care of nature, nature will take care of us. Conservation has truly come of age when it acknowledges that if we want to save part of the system, we have to save the system itself. This is the essence of what we call sustainable development.'

Ultimately, people in Scotland, and every other nation, will have to understand environmental issues in world terms. International reports continue to criticise governments for spending billions of pounds on the destruction of land, oceans and the atmosphere. The authoritative Worldwatch Institute in Washington estimates that world governments are spending: $100 billion a year subsidising power stations which increase global warming; $300 billion encouraging overgrazing and destructive farming; and $50 billion on over-fishing. In its State of the World 1997 report, Worldwatch said that since the Earth Summit in Rio in 1992, the world population had grown by 450 million (more than the combined population of Russia and the USA); emissions of carbon dioxide had risen to a new high; and overseas assistance from the richest countries to the poorest had fallen to its lowest level since 1973 – just 0.3 per cent of GNP. It also highlighted the inescapable fact that the developed world bears primary responsibility for the planet's ills. While each citizen of America leads a lifestyle responsible for 5.3 tonnes of carbon dioxide (CO_2) a year, each Indian is responsible for only 0.3 tonnes of the global warming gas.

Meanwhile, the UN Environment Agency in Nairobi has calculated that three billion people will be short of water within fifty years; that 1.23 billion acres of African land – twelve times larger than the area of Britain – have suffered moderate to severe soil erosion; that most oceans are being over-fished; and that three-quarters of the world's species are declining or facing extinction.

Finally, a UK government report found that Britain was spending £20 billion a year on environmentally damaging industry, energy use and agricultural grants. It concluded that a new philosophy was needed to avoid crippling social and economic decline.

I can't help thinking that Fraser Darling would not be sanguine about the progress made since he described Scotland's wet deserts. But he might welcome the conviction of a new generation that their fathers, grandfathers and great-grandfathers may have got it wrong.

2

·

Uncommon Land

·

The ground is holy, being even as it came from the Creator.
Keep it, guard it, care for it, for it keeps men, guards men, cares
for men. Destroy it and man is destroyed.

Alan Paton, Cry the Beloved Country

Mist and Magic

·

THERE IS A WORLD of mist and magic in Scotland, which is
worth preserving and understanding. But first we have
to save it. I have experienced the joy of something like
wilderness in the restored native Caledonian pine forest
at Abernethy, a site owned by the Royal Society for the
Protection of Birds (RSPB), and on the protected islands of
Loch Maree. I have imagined the old woods and the old
landscapes, in the bones of real pine forest exposed on the

Mar Lodge and Glenfeshie estates in the Cairngorms. I am not hopelessly romantic about this. I once spent several days sleeping in a hammock strung between trees in an area of primary tropical forest in Brazil. And despite the dozens of species of poisonous tree frog I witnessed at close quarters, the tarantula with which I had a tug-of-war (with a twig), and the alligator swimming in our watering hole, I hated it. I sweated and suffered, was depressed by the absence of light, and understood the urge to reach for a chainsaw and create my own space. I understood, also, why I should not. We often forget that in Brazil, Malaysia and Indonesia we are still destroying real wilderness. In Scotland, we have to wonder at the small fragments of natural and semi-natural forest remaining, and try to expand them.

Of course, the tropical forest bears no resemblance in its extraordinary richness and character to the temperate Great Wood of Caledon, which once marched from Loch Lomond to Loch Shin, from the Braes of Angus west to Ardgour. That wood was a home to bear, elk, lynx and wolf, but, even in the mists of time, could not compete with the myriad species of trees, flowering plants and insects found in the tropics. Yet, like the modern ranchers, landless peasants and developers in Brazil, the first farmers in Scotland saw the forest as a threat and a hindrance to the lifestyle they desired. So they cut it and burnt it. I believe that if we are to pass judgement, and express concern, about what is happening in the tropical rainforests of the world today, we must first deal with our own degraded landscape.

We cannot recreate the Great Wood of Caledon. But we can protect and enhance native woodland where it exists – which is more than we have done for most of this century – and we can expand woodlands in rural areas for social, economic and environmental gain, and to create a more natural Scotland. We need not destroy farming to do so, we need not eliminate sporting estates, but we should aim for a greater mix of species and habitats in our uplands, which have for too long been dominated by extractive mono-cultures of sheep, deer, trees or grouse.

Dick Balharry, the retired Scottish National Heritage

(SNH) conservationist, once told me: 'I blame society. Society has got to make up its mind whether these treasures are important enough to protect them in perpetuity.'

He was quite correct. The history of the destruction of these woods, is the history of Scotland; their future, Scotland's future. We need to venerate the ancient sentinels which have survived; trees like the lone pine tree at Arkaig, Inverness-shire, which was 150 years old at the time of the last Jacobite rebellion in 1745. In a natural environment, we would have many thousand such trees, creating a more diverse environment, containing many more species than our uplands currently support.

On a world scale, preserving biodiversity means preserving the potential discovery of important medicines, pharmaceuticals, timber, fibre, soil restoring vegetation, petroleum substitutes and countless other products. Worldwide, around 1.82 million species of living things have been identified, of which 1.04 million are insects, 325 000 are plants and 41 000 are invertebrates. There may, however, be as many as 100 million species currently in existence, at a time when we are losing species at the rate of three an hour, or 27 000 a year. We are living through the fastest period of extinction.

Scotland is a part of that shrinking natural world, its responsibilities every bit as important as those facing any other European nation. We have some of the finest landscape and wildlife resources in Western Europe, but they lack the level of protection which the wider conservation community deems necessary. In 1995, the Parks for Life taskforce of the IUCN described the management of the Cairngorms and Loch Lomond as inadequate.

One of the maxims of the Sierra Club, a venerable American environmental organisation, is that wilderness will be preserved in proportion to the number of people who know its values at first hand. The maxim bodes ill for Scotland. We are, as a rule, ignorant of the natural world around us. Here, as in many developed and developing nations, there has been a pattern of movement, of people and resources, from upland and mountain areas to lowland areas. Planners, politicians and the public have habituated the current

Scottish landscape and accepted it as natural. Since the
1950s, traditional knowledge of the natural world has been
lost with successive generations.

Yet even a cursory examination of the different states of
vegetation inside and outside a simple stock fence – along
the edge of the A9 in Drumochter, for example – can
illustrate the unnatural poverty of moorland kept plain by
grazing animals. Inside the fence that protects the railway
line, there are trees, shrubs, insects and birds. Outside, there
is short heather. The example is simplistic, but helps to
illustrate the absence of any integrated land management
strategy. In our most important wild land areas, the conservation mechanisms that exist are not strong enough to
safeguard habitats and species. Instead, the health of great
tracts of land is left to the piecemeal efforts of landowners,
planners, government agencies and environment groups.
The availability of inappropriate agricultural subsidies,
which keep too many sheep on the land, and the laird's pre-
occupation with grouse and deer, do little to help.

Scottish Wildlife and Countryside Link, the umbrella
body for Scotland's environment and recreation groups, put
it this way: 'Scotland's finest landscapes are unnecessarily
blighted by inappropriate forestry, the construction of
hill roads, poorly planned or designed energy generation,
quarrying, mining, tourism, ski developments, holiday
houses and other buildings.'

For too long, the management of wild areas in Scotland
has depended on the much-derided 'voluntary principle',
under which SNH seeks to persuade landowners to manage
the land in a positive way, and even pays them large sums of
money not to damage designated sites.

In the early 1990s, the Conservative Government, with
SNH, tried various partnership approaches to the management of the Cairngorms, an area regarded as one of the
most important wild land areas in Europe. They made
significant advances, but ultimately failed to secure the
natural regeneration of native woodland. This should not
be an option in a civilised society, but should be provided by
regulation and legislation. It is both remarkable and inde-

fensible that the Cairngorms, the jewel in the crown, should be listed for World Heritage Area status – St Kilda is the only natural environment which has the accolade in Scotland – but should be unable to achieve it because the mountains are degraded.

Getting things right in future may involve looking at Scotland in a different way. Ron Greer and Derek Pretswell, of Natural Resources Scotland, are former fisheries scientists who are fond of turning their country on its head. They like to turn the atlas upside down and look at Scotland from the Arctic. It looks different.

Life in Tromsö, inside the Arctic Circle, has the same basic climatic recipe as life in the Cairngorm ski centre car park, but with different results. In the Norwegian university town, there are hills clad in birch, elm trees in the street, and warm houses. In the car park above Loch Morlich, there are rocks, coarse grass and some heather.

From an Arctic perspective, we have a good northern climate, rather than a bad southern climate. Bio-climatically, many areas of the Highlands can be closely matched to parts of Norway, yet in areas like the Drumochter Pass, where we have the wet heather deserts derided by Fraser Darling, the Norwegians have successful agricultural and forest communities and enjoy a high, sustainable, standard of living.

With approximately the same ingredients, we create monocultures and the Norwegians have orchards, varied wildlife, timber production, sheep farming and community ownership.

The point was well made by Angus McHattie, a Skye crofter, who wrote: 'On returning from Norway to Skye recently, I had occasion to compare the view from similar 3000ft granite hills in both countries. In Norway, the valley I looked down upon contained an autonomous village of 20 small farms, with their own crops, power supply, school, etc. – a prosperous and happy place with a good trade surplus and a population with a healthy age structure. The Skye valley had 20 blackface ewes and 12 lambs.'

Greer and Pretswell have, for some time, wanted to take 50 000 acres on either side of the A9 and transform it into something like the Norwegian model. They want to re-introduce people, trees, extinct herbivores – like the aurochs (wild cattle) – and businesses based on the managed stock and forest products. If the project succeeded, the resulting landscape would be a model of land management, and would prove the potential of the unending moors.

Their ideals also illustrate one of the other factors limiting development in the Highlands: a system of land tenure in which individuals, usually absentee owners living in the south of England or foreign nationals, own huge tracts of land and manage them without reference to the local community and often without any holistic approach to the natural environment.

In Rogaland in western Norway, the population rose from 211 000 in 1952 to 309 000 in 1981. In Perthshire, which is seven times larger, the population fell over the same period from 128 000 to 119 000. There are 10 000 landowners in Rogaland, while just 800 individuals own most of Scotland.

The Norwegian farmer owns his house, a paddock, part of the woodland, and shares with the community the ownership of upland grazing. Sheep are in-wintered, reducing mortality and leaving the shepherd free to diversify. In upland Perthshire, the tenant farmer does not own the land, has no access to the forest and keeps his sheep out of doors.

The issue of land ownership has been dealt with elsewhere, but it is encouraging that the Labour Government plans to end the ancient feudal system under which a landowner can wield power over home owners long after having sold the land on which the house is built. The Government has also set up a unit within Highlands and Islands Enterprise to study the potential for community-led land purchases, and the Commons select committee on Scottish affairs has been asked to look at the 'land question'.

Duncan Macrae, one of the crofters instrumental in the purchase of the North Lochinver estate by the Assynt crofters, has noted that it is good to wake up in the morning and know that what you see from your window belongs to you. In most Highland villages, that feeling does not exist. Macrae's appreciation of the view from his window hints at the bond between man and the land which was celebrated by Gaelic poets and is promoted today by the Wilderness Trust, an organisation which takes people into wild areas to reintroduce them to nature – in the African bush, on a remote Canadian peak, or even in the hills of Knoydart. The trust believes that people have lost the spiritual connection to the natural world which evokes concern for the environment, and other fine virtues. The Highland landscape is hardly natural, but many people who live there still feel strongly attached to the land, and visitors may discover for themselves an unlooked-for spirituality in their contemplation of the hills and glens.

Some Issues in Detail

The over-population of red deer is one of the most readily solvable problems facing upland Scotland. High deer numbers are supported by bad management and prevent the growth of young trees, unless the grazers are fenced out. But deer fences are not the answer because they create thickets, not natural woodland, and because they cause the death of large numbers of important game birds, such as capercaillie.

The deer problem is particularly one of hind numbers, and despite record culls of more than 60 000 stags, hinds and calves in recent years, the overall population remains stable. Some landowners want large deer populations because the value of the sporting estate is calculated according to the number of stags shot, while others say they want to reduce the herds, but lack the manpower for the task. Either way, it is bad management.

Until recently, the Deer Commission for Scotland colluded in allowing the red deer population to grow to around 350 000 animals. Following its reorganisation in 1997, it was given responsibility for the habitat, as well as the species. This gave it the option of shooting deer out of season, and of sending in its own stalkers to do the job properly. Owners and stalkers set great store by tradition, but solving the crisis in some areas may require the use of helicopters, which can spot the deer, drop the stalkers near to them, and then pick up the carcasses.

Where major reductions have been achieved, the effects are dramatic. On the RSPB's Abernethy estate, there has been a 25 per cent increase in tree seedling and sapling numbers over three years in response to a 40 per cent cut in deer numbers. Similarly, there is healthy regeneration taking place on Beinn Eighe and Creag Meagaidh, which are both owned by SNH.

Other issues, particularly those caused unwittingly by man's activities, have less obvious solutions. Acidification, in particular the deposition of atmospheric nitrogen, is altering the vegetation on our hills. After carbon, hydrogen and oxygen, nitrogen is the most abundant element in plant tissues and its supply often limits plant growth. Therefore, an increase in the deposition of nitrogen affects the most sensitive ecological systems, which have low nitrogen availability in the soil, and contain plants adapted to low nitrogen supply. One such moss, *Racomitrium lanuginosum*, is the most extensive of all near-natural plant communities in Britain. It has no roots and takes its nutrient supply from the atmosphere. This heath has been deteriorating south of the Highlands for more than sixty years, with a corresponding increase in grasses. The loss of the moss and its replacement by grass is often a result of the combined effects of added nitrates and the grazing/trampling pressure of sheep.

In an experiment in the early 1990s, *R. lanuginosum* was taken from Rannoch Moor and transplanted in the Pennines upwind of Liverpool and Manchester. It showed a progressive increase in nitrogen content over a fifteen-month period, illustrating a process which has already had a damaging effect in the Pennines, the hills of Galloway and elsewhere south of the Highland line. The effect of such changes is one of the most important conservation issues in the uplands, which account for 70 per cent of Scotland's land surface, and is closely linked to increasing emissions of nitrogen from vehicle traffic.

Since the 1940s, the total extent of upland semi-natural habitat types, like heather moorland, has been reduced by

12 per cent, largely due to the expansion of commercial forestry and conversion to grassland. A much larger area, possibly 30 per cent, may have been degraded by grazing, bad muirburn practice and acidification. More generally, there has been a widespread decline in the abundance of heather over the last 150–200 years.

The EU's habitats directive is leading to the designation of Special Areas of Conservation (SACs) in Scotland, but the task is daunting. Des Thompson, an SNH upland expert, suggests, for example, that for woodland and scrub communities the management aims in an SAC should include: maintaining existing semi-natural woodlands; enhancing them through favourable management; extending woodland cover through the expansion of existing woodland remnants; and recreating subalpine and tree-line scrub on a series of sites. Such immodest targets also apply to areas of mountain moss and lichen, blanket bog, dwarf-shrub habitats, and so on.

The Making of our Natural Heritage

Geology

Much of central Scotland is made up of a hard crystalline core of schists and granites, while the gentler, more fertile lowlands and the Borders are composed of softer, sedimentary rocks. Volcanic rocks form most of the hills in the lowlands, including the Ochils, the Sidlaws and the Campsie Fells.

In its own relatively small area, Scotland exhibits an incredible geological diversity which attracts students of science from around the world. There are rocks from most periods of geological time, including some of the oldest rocks in the world, such as the banded gneisses which make the lochan-dotted scenery of much of Lewis and Harris in the Western Isles. These are more than 3000 million years old, while the sedimentary rocks of the midland valley in central Scotland are a mere 300 million years old.

The ancient gneisses predate numerous periods of

mountain-building, volcanic activity, many ice ages and the splitting-up of the continents. The importance of Scotland's geology is indicated by the fact that around one third of the 1300 protected Sites of Special Scientific Interest (SSSIs) are designated for their geology.

Although the detailed geology is complex, there is a basic structure which can be divided into five broad zones of distinctive landforms and soils:

1. The north-west Highlands and Islands west of the Moine Thrust – composed of ancient Lewisian gneiss.
2. The north Highlands between the Great Glen Fault and the Moine Thrust – the eroded roots of the Caledonian mountain belt, primarily made of metamorphosed sediments laid down 1000 million years ago.
3. The central Highlands between the Great Glen Fault and the Highland Boundary Fault – composed of 800-million-year-old Dalradian sediments, with 'intruded' younger granites forming the Cairngorms and Glen Coe.
4. The Midland Valley – home to most Scots – made of sediments deposited in tropical seas 300 million years ago, and the lava fields of the Ochils and Clyde Plateau and the volcanic rocks of Edinburgh and Fife and East Lothian. The rocks here include coal and limestone.
5. The Southern Uplands, composed of sediments deposited in the Iapetus Ocean, which, for more than 100 million years, separated Scotland and England.

Laid on these ancient landscapes are the effects of successive ice ages, the most recent of which scraped much of the soil from the area of old, hard rocks north of the Highland boundary fault.

Scotland's geological heritage has also relinquished fossils which provide an insight into the beginnings of terrestrial plant and insect life. A chert deposit, near the village of Rhynie in north-east Scotland, has produced some of the first plants recorded in the world, and the oldest known fossil insect. Primitive fish have also been found in deposits in central and northern Scotland, and 'Lizzie the Lizard', which may be the earliest known fossil reptile, was uncovered in a disused quarry in Bathgate. Geology also determines the mineral wealth of a nation. Early man collected flint in glacial deposits and today coal is mined, and stone and aggregate are quarried for building.

Climate

Scotland's position on the west coast of Europe, in an oceanic zone, produces moderate temperatures for a latitude which is level with Moscow. The average temperature of the coldest month in Lerwick is 38 C, compared to minus 108 C in the Russian capital, which is 500km further south. Our climate is dominated by the ameliorating effect of the Gulf Stream, and a predominantly westerly airflow which brings rain from the Atlantic. The mainland is only 240km at its widest, but there are dramatic differences in rainfall from east to west.

Fort William receives 1400mm of rain every year, twice as much as Glasgow, and four times as much as Edinburgh. On western mountain summits, the rainfall can reach 3800mm, while in the drier parts of the east coast – the East Neuk of Fife, for example – it may be as low as 600mm.

The warmest part of Scotland in mid-summer is not Fife or East Lothian, but Glasgow and urban central Scotland, including Falkirk, Bathgate and Linlithgow. The area benefits from the fact that it is geographically central in the Scottish landmass and low-lying. However, the climate probably feels better in East Lothian, which has similar temperatures, and less rain.

Relict Geomorphology

The landscape today has an unusual number of forms created under conditions which no longer occur. These are called 'relict' landscapes and are vital tools in working out bygone environmental conditions. Many sediments in lochs and bogs, for example, are living records of environmental change. Large-scale relict forms included the corries, glens and loch basins created by glaciers in the Highlands, while much of lowland Scotland is not much different from its pre-glacial state.

Dynamic Geomorphology

The Scottish landscape is still changing. Some changes, such as the loss of sand dunes, the build-up of sediments in rivers, or the creation of gullies on hills, can take place over a short period, while others, such as the weathering and wind abrasion of rocks, are so slow as to be imperceptible. Another very slow process is the rebounding of coastal land, which is still rising following the departure of the last ice cap.

Man-made impacts on geomorphology vary greatly. In some areas the construction of coastal defence systems to prevent erosion can damage and even destroy neighbouring sand dune systems.

Winter floods, like the one in 1993 that devastated large areas of farmland and flooded hundreds of homes in the Tay valley, are examples of the dynamic behaviour of river systems. On the Tay, the construction of embankments, to allow the cultivation of low-lying fields, served to usher the flood downstream to the housing estates of Perth.

Soils

Soil is the foundation on which every natural environment is built, but its health has often been neglected in Scotland. Soils are the interface between rocks and biology, and where they are damaged, whether in farming areas or forestry plantations, biodiversity declines. Soils reflect the processes from which they were formed, and the influences of human activities, present and past. The point was nicely illustrated by a study in the USA in which bore hole samples taken from graveyards proved to be healthier and richer than the soils in the surrounding, altered landscapes.

Soils are made up of fine rock material, sand, silt and clay, together with organic matter – decaying plants and animals – and air and water. They are reservoirs of nutrients, water and minerals, Once damaged, they are difficult to restore.

Soil Development

Scottish soils are very young in geological terms, having developed since the end of the last Ice Age, around 10 000 to 15 000 years ago. The retreating ice sheets left behind a strange, barren landscape of glacial drifts, including rock and fine-grained material, which was enriched slowly by the effects of physical and chemical weathering and the breakdown of plants. Over thousands of years, the soil developed a definite structure and was able to retain moisture and gather nutrients and organic material. The only areas where the process did not take place were the mountain-tops and the most exposed coasts, where primitive soils remain.

Somewhat surprisingly, it was 1996 before a UK government recommended that soils should be given the same protection as air and water.

Scottish Soil Types
Drift soils, in which local materials have been deposited by glaciers, are found over 80 per cent of Scotland. Organic soils, like peat, are the next most extensive type, covering 10 per cent of the country. Most Scottish soils are naturally acidic. In general, the most acid soils are in the uplands, and the least acidic in the lowlands.

There are four basic types – peats, gleys, podzols and brown forest soils – all of which are influenced by changes in one of the five determining factors of soil formation. In the broadest terms, the uplands are dominated by peaty soils which have been altered by the management of sheep, deer, forestry and sporting interests. The lowlands are mainly mineral soils, often changed by agriculture, forest clearance, drainage and the application of fertilisers.

Over much of the Western Isles, northern Scotland and the Highlands, wet and cold conditions have suppressed biological activity, allowing plant debris to accumulate faster than it decomposes. The result is peat.

At lower levels, with lower rainfall and higher temperature, there is more micro-organism activity which promotes the development of better soils, although still with organic material on the surface. These 'gleys' develop on fine-textured parent material with poor drainage. Where air is excluded from the soil, iron and other minerals are reduced due to lack of oxygen, and the sub-surface colours of the soils are grey, blue and green, with rusty brown colours where oxygen reaches into the soil – as in Ayrshire and Lanarkshire.

In the drier east of Scotland, the parent materials were often acidic and coarse grained, and podzols dominate. In these, acid conditions have suppressed biological activity, allowing organic accumulation on the surface. But their coarse sandy texture means they are well drained and minerals are leached out of the upper layers. This results in a profile of different layers, with dark, surface organic material, a coarse grained, bleached upper horizon and a dark brown lower level where the minerals are re-deposited.

In the south-west of Scotland, peaty and gley-type soils dominate. In the south-east, where conditions are drier, brown forest soils have developed. They are formed in non-acid conditions, with low rainfall and warm temperatures, all of which encourage the biological activity which efficiently breaks down organic matter to produce deep, fertile soils with large quantities of earthworms. They are the best agricultural soils.

Some of the earliest farmers practised soil enrichment, spreading human and animal manure and seaweed on the poor soils of the Highlands.

Erosion

Erosion is a natural process caused by water and wind which supports soil formation, but can also be induced by man's activities. Causes of 'fast' soil erosion include:

- overgrazing by sheep and deer;
- the seasonal burning of moorland, especially heather, for the benefit of grouse and sheep;
- the removal of hedges and other windbreaks;
- downhill ploughing;
- winter cropping, which leaves soils exposed to high rainfall and strong winds;
- deep ploughing and drainage for afforestation; and
- the expansion of path networks.

Erosion is difficult to assess and current knowledge of the problem is limited, but mineral and peaty soils in the uplands are known to be eroding. Because soil formation is such a slow process, the effects of erosion, which depletes organic matter and reduces soil depth, are very difficult to counter. At many lowland sites in eastern Scotland, accelerated erosion has been recorded in winter. Studies in the 1990s showed that soil was being lost in the Kelso area at a rate of up to 80 tonnes per hectare per year, compared to an accepted rate of loss of one tonne. A survey in the Earn Valley showed that the most eroded fields were under cereal crops sown in the autumn, or were ploughed. And in Shetland, the rate of soil erosion was recorded at between one and four centimetres a year on bare peat surfaces, which could result in total loss of peat in some areas within 30 to 150 years.

Pollution

We are still learning about the effects of pollutants on soils. The main sources are local dumping – for example, colliery spoil in central Scotland – and the much more extensive airborne pollution, which can bring chemical pollutants from as far as Poland and Germany.

Acidification is a major threat and a large area of sensitive soils will continue to receive excess acid deposition beyond 2005 unless current pollution emission rates are reduced substantially (see Chapters 4 and 6). Acid deposition is exacerbated by pollution in areas of high rainfall and low temperature with freely draining, acid soils.

Many areas of Scotland are vulnerable to acid deposition in Scotland. The north-west of the country is little affected because the pollution, much of it from the big coal-fired power stations of Yorkshire and the Midlands, has already been deposited further south, in Dumfries and Galloway and the Trossachs.

Land Cover

There are few unaltered landscapes in Scotland. We get nearest to a 6000-year-old landscape on small, densely vegetated islands, isolated peatlands, cliff faces and rarely visited mountain summits.

A survey which compared the land cover of Scotland in 1946, 1964 and 1988 showed a tenfold increase in built-up areas and a ninefold increase in coniferous plantation. Heather and moorland, covering a third of the land area in 1988, had declined by 15 per cent since 1946, and the overall pattern was of increasing disturbance and modification, mainly associated with afforestation and the reduction of semi-natural land cover. Long established or semi-natural woodland declined in the same period because of the under-planting – no longer practised – of commercial conifers such as sitka spruce and larch. The total length of lines of trees on roadsides and field margins has decreased by 9 per cent. Change is least obvious over time in the agricultural lowlands of Fife, Lothian and Orkney, where there is little semi-natural land cover.

A study in 1990 found species diversity had decreased in improved grassland areas, and the number of woodland species was down everywhere, while woods had become more open and grassy. The length of hedgerows, dominated by hawthorn, was found to have dropped from 37 000km to 33 000km between 1984 and 1990. Animal diversity was greatest in lowland fresh water, although water quality was highest in the uplands.

Biogeography

The distribution of animals and plants – and the look of the landscape – is influenced by environmental conditions including geology, soils, landform, exposure and climate.

Because Scotland has a wide diversity of landforms and climates, it has a similar diversity of plant and animal communities which have adapted to mild, wet maritime conditions on the west coast, and the Arctic–Alpine conditions of the high mountain plateaus and summits.

All species have geographical ranges, although some, like the killer whale, have a global distribution. Scotland is at the northerly limit for its eight species of bats, while the Scottish primrose is restricted to damp, cool, exposed northern coasts. The snow bunting confines itself to mountain boulder fields, and the ptarmigan stays on heather moor.

Landscape

Scotland's world-famous landscapes range from the wide, low-lying glaciated valleys of the north west to the flat, open landscape of the northern Flow Country, and the sloping, arable farmlands and smooth coastline of the north-east. There are jagged peaks in the Black Cuillins of Skye, mountain ranges in Kintail and Glencoe, and single mountains such as Ben Assynt. The whole lot is given an aesthetic dimension by the interplay of air, water and light, and the changing weather and tidal conditions of Scotland's North Atlantic seaboard.

Human activity, however, has been redefining the look for 6000 years. The hilltops of Glencoe will hardly have altered in that time, but the glen, which was once forested and a home to large carnivores, is now bereft of trees.

Cultural changes in the landscape have been moving apace in the last 300 years, with the growth of cities following the industrial revolution. Many settlements in our countryside grew up because they were linked to water crossings, good harbours or fertile soils, but today their *raison d'être* is more likely to be fish farming or tourism.

In the uplands, land management for farming, forestry and field sports has influenced the creation and modification of moorland. In the west and north, crofting and fishing communities colonised the rocky coasts after landowners instituted the eighteenth- and nineteenth-century Clearances in which they removed people from their land to replace them with sheep.

As recently as the past two decades, leisure and tourism activities have had an impact on the Scottish landscape, with the development of ski areas and loch-based water sports.

The built environment can enhance the view when natural and local materials are used – the grey granite houses of the north-east are an example – but the Highlands are, in many places, blighted by modern box bungalows and ugly council housing, which, as the Prince of Wales might have said, are like a rash on a beautiful face.

Mountains

Scotland's mountains cover 12 per cent of the country and are defined as land above 700m, the approximate equivalent of the former tree line, which has been lost almost everywhere. In the north-west, the tree line drops to 550m and in the Northern Isles to only 200m. The relatively low tree line in this country reflects the oceanic climate; in the Alps it reaches 2600m, and even in Arctic Norway an impressive 500m. Mountain habitats are mainly made up of moss and lichen heaths, snow beds, blanket bog and dwarf-shrub heath. The internationally important features of these high areas include vegetation that is globally rare and breeding birds from the Arctic.

In a UK context, the Highlands are important because they contain the greatest amount of near-natural vegetation. In the montane SSSIs, 87 per cent of the vegetation is semi-natural, rather than anthropogenic. The greatest diversity of vegetation occurs where the determining factors – climate, geology, etc. – are most varied. Good examples are the Cairngorms and Ben Lawers.

Mountain habitats are threatened by sheep and deer grazing, recreation, airborne acid pollution and, more insidiously, global warming. An increase in average temperatures of 1 8 C would be the equivalent of shifting habitat zones 200m higher.

Moorland

There are thirteen heath and thirty-nine grassland types in Scotland, which are valued by naturalists for their plant and bird life, and by historians and archaeologists for their historic and prehistoric landscapes They are used for grazing by sheep and deer and as managed grouse moors. Their condition is determined by burning for grouse and sheep, and grazing pressure. Much modern moorland, and most grasslands, have been created by management over several centuries. If grazing pressure was reduced, large areas would revert to scrub and woodland. The main losses have been due to forestry, bracken and sheep grazing which have caused heather to convert to grassland. Monitoring suggests that between the 1940s and 1970s, five regions – the Northern Isles, Grampian, Lothian, the Borders, and Dumfries and Galloway – lost 29 per cent of their dry heather moor. Many of Scotland's familiar open moors were formerly covered by woodland, but in some cases the woods disappeared, because of climate change, thousands of years ago.

Burning the land to improve the pasture dates back to Neolithic times, although muirburn, in which the vegetation is burned early in the year for the benefit of sheep or grouse has only been common since the 1840s. Today, it is often a damaging practice. It has a serious impact, for example, in the wet heaths and blanket bogs of north and western Scotland, where shepherds burn the heather with little control to provide an 'early bite' for sheep. In this area, the heather cycle is twice as long as in the east and the thinner, peaty soils are more susceptible to rain and wind erosion after burning. Some upland scientists question the need for any burning in the west. In South Harris, centuries of burning and over-grazing have wiped out dwarf juniper and heather, and helped to expose bare rock.

On the other hand, when muirburn is carried out well on the drier, eastern grouse moors, it can maintain the mosaics of young, intermediate and old heather which hold a wide variety of invertebrates and birds. The new growth after fire provides shoots for feeding grouse, and the birds can escape from predators like the peregrine falcon and golden eagle into mature heather.

Nonetheless, the number of grouse shot each year has fallen to half its 1935 level, probably due to a mix of factors including increased predation from adjacent plantations, loss of heather, a reduced level of game keepering, competition with sheep, bad muirburn practice and disease.

Around forty bird species are found on moorland, including twenty-one that breed on the moor. But only red grouse and golden plover benefit from muirburn. At least five species, including the hen harrier, the oyster-catcher and the snipe, are declining in numbers.

Peatlands

Peat is a major, and important, habitat type for Scotland. It is an organic deposit made when dead plant material decays in waterlogged conditions; peatlands can be up to 98 per cent water by weight, which is more than a silt-laden river. It begins to form when nutrients have been flushed out of the soil by a wet climate, leaving the ground waterlogged and infertile. These conditions suit sphagnum, which needs little nourishment and builds up at a rate of around 1mm a year. Eventually, the original mineral soil is left below several metres of peat. Scotland's 'blanket' peat bogs cover 14 per cent of the countryside, and are more common here, and in Ireland, than anywhere else in Europe. In lowland areas peat tends to form as isolated domes called raised bogs. Bogs are species-poor, but support a variety of unusual and specialised plants and animals, including the insectivorous sundew which eats midges. They are also natural archives which have stored pollen records for up to 7000 years, and have yielded well-preserved human artefacts and remains. Scotland has 70 per cent of Britain's bogs.

Raised bogs have been reduced to 11 per cent of their former extent, and are found in the Central Belt, the coastal area of Grampian and along the Solway. They have been damaged over the centuries by reclamation for agriculture, by planting – trees dry out the peat – and by drainage. The total area of raised bog dropped by 23 per cent between 1940 and 1970, mostly due to afforestation, but the main concerns at present are open cast mining and peat extraction for horticulture.

Blanket bogs dominate large areas of the Scottish uplands and extend to sea level in the far north and west. The Flow Country in Caithness and Sutherland is probably the largest single expanse of blanket bog in the world, with a wide range of breeding birds, particularly waders. Until the late 1980s, large areas of the Flow Country were being planted by landowners for tax reasons. The environmental outcry that followed resulted in the planting incentives offered by the Forestry Commission being changed, and agreements being reached on the preservation of the habitat.

Forests and Woodlands

The native forests that once covered 80 per cent of the UK have been altered and destroyed by man for thousands of years. Some 2000 years ago, forest cover had been reduced to 50 per cent of the land area; it had dropped to below 20 per cent by 1000 AD. In the fourteenth century, it was down to 10 per cent, and by the beginning of this century, trees accounted for just 5 per cent of the land area.

Following the First World War, the Forestry Commission was established with the aim of making the UK self-sufficient in timber. It was a hopeless target. After eighty years of planting, the forest area has expanded to cover 14 per cent of Scotland, but the UK still imports 80 per cent of its timber.

However, conditions for forest growth are highly favourable. Scotland sits at a boundary between the 'natural boreal coniferous' and the 'broadleaved deciduous forest' regions. Unfortunately, most of the forest area is dominated by imported species such as sitka spruce and larch which were chosen for their fast growth. Native species – Scots pine, oak, birch, etc. – account for just 17 per cent of the total forest area, or less than 2 per cent of the land surface. The commercial timber market in Britain is dominated by exotic species, although in the longer term there is no reason why native species should not be used in paper and board mills, as they are in Sweden and Norway.

Our planting history is also idiosyncratic for the fact that between 1945 and 1986, agriculture enjoyed a greater priority than forestry, so most plantations were created on poor upland soils. And, until the recent reorganisation of priorities within the Forestry Commission, and the development of an environment ethic, planting had many negative impacts.

As recently as the 1980s, commercial plantations were usually single-age, single-species blocks, dumped with little imagination into the upland landscape. The afforestation of upland grazings and peatland destroyed reservoirs of wildlife, and, more locally, large coastal sand dunes – at Tentsmuir in Fife, and Culbin in Morayshire – were afforested and lost their conservation value.

The RSPB said the planting of the Flow Country between 1981 and 1988 was carried out with 'little consideration for forestry merit, financial appraisal or environmental impact'. Of the seventy-one bird species found on the uplands, it is thought that thirty-four are at some risk from afforestation, with substantial declines recorded for fourteen species.

But new guidelines have changed all sorts of bad practices, and an element of native woodland is required in every new planting scheme. Sites have to be considered for their impact on the landscape, and areas of low ecological interest have to be enhanced. New planting is not allowed to endanger important species and habitats, and cultivation and drainage have to be designed to minimise erosion and any potentially damaging effect on watercourses. Modern design standards impose strict conditions on drainage gradients, and the spawning areas of salmon have to be protected.

The green revolution has been so great that some conservation bodies now describe the Forestry Commission – which has been divided into Forest Enterprise and the Forest Authority, the commercial and regulatory arms – as the greenest government agency by far. It is also working quite effectively to increase native woodland cover through attractive grants for the creation and enhancement of native pine, birch and oak woodlands.

Of course, this is not before time. Since the creation of the Commission in 1919, the total area of semi-natural woodland, including Scotland's ancient woods, has been in decline. While the area under plantation increased threefold in the three decades after 1940, the area of native woodland dwindled by 15 per cent. The clearance of native species has also removed all of the natural tree line, with one remarkable exception, at Creag Fhiaclach, at 640m in the Cairngorms.

The surviving remnants of ancient woodland – there are eighty-five Caledonian pine areas – are the vestiges of the natural 'climax vegetation' which covered the country at its biological peak. These include the ancient oak woods at Taynish and Loch Sunart on the west coast, which are recognised internationally as important remnants of temperate rainforest.

The Caledonian woods of pine and birch, which can still be seen at Loch Maree, in Glen Affric and at Mar Lodge, have a smaller range of species than other continental boreal forests because of their island setting and oceanic climate. They include rare plants such as twin-flowered and one-flowered wintergreen and are home to the Scottish crossbill, Scotland's only endemic bird species.

The extent of the pinewoods has fallen by roughly 25 per cent over the past thirty years, and conservation measures are just beginning to address the issue. The Commission first made funding available to save and extend these woods in 1978. The situation will be further improved by its new objectives, which aim to promote the economic, social and environmental benefits of new forests.

The current trend is for most planting to be carried out by private companies and landowners taking advantage of woodland grants, and the area of land planted annually by the commission itself has been falling steadily since 1970. The existing forest resource is divided roughly equally between the public and private sectors.

Woodland cover is also being affected to some extent by the work of environmental charities. The Central Scotland Countryside Trust has been working towards the creation of a new forest between Edinburgh and Glasgow for several years. It aims to create a multi-purpose mosaic of woodland, farms, towns and business sites, to encourage economic regeneration and to enhance the quality of life in a landscape ravaged by old industrial developments, such as mining for shale and coal. There is a target of 17000ha of new woodland by 2015, although the imbalance between agriculture and forestry grants – the former are more attractive – and the reluctance of some farmers to put land into woodland was hampering progress.

The Millennium Forest is another ambitious scheme which aims to restore and regenerate native woodland throughout Scotland, in towns and cities, and in remote parts of the Highlands, wherever there are environmental and social benefits. The project, which encourages community involvement, has been awarded over £11m in Lottery funding since 1995, and many individual schemes are underway. Some of the money has gone to native pinewood restoration work in Glen Affric, where the commission is working towards the same goals as the National Trust for Scotland and the charity Trees for Life.

3
·
Wildlife
·

Wild life does not exist for man's delectation. Man may find it beautiful, edifying, amusing, useful and all the rest of it, but that is not why it is there, nor is that a good enough reason for our allowing it to remain. Let us give beast and bird and flower the place to live in its own right.

Sir Frank Fraser Darling,
Natural History in the Highland and Islands

The Return of the Wild
·

THE WHITE-TAILED SEA EAGLE fixed me with its black eyes and shifted, almost imperceptibly, on its perch. A touch ambitiously, I tried to mirror the gesture. Uppermost in my mind was a desire not to make any sudden movement in the confines of the cage. I was alone with Britain's biggest bird of prey and the world's fourth largest eagle. Neither of us appeared to be quite at ease.

The female bird had not yet developed its distinctive white tail – that would happen in a year's time – but at the age of just fourteen weeks she was already fully grown, with a formidable black beak, massive talons and an eight-foot wingspan. In Gaelic the bird is known poetically as *iolair suil na grein*, the eagle with the sunlit eye. The affectionate name not withstanding, it was persecuted to extinction in the 1910s.

Now, it is now back where it belongs, riding the thermals over cliff and coastline from Rum to Dunnet Head. A few weeks after my uncomprehending encounter with the young bird, it was released with another nine, from cages hidden on a west Highland estate, to join a population of more than fifty sea eagles surviving in the wild.

Like the others released in recent years, it had been taken from a nest in northern Norway, flown to Scotland, and kept in a cage for eight weeks while it was fed deer, goat, rabbit and hare.

In preparation for freedom, the bird was tagged, measured and weighed, and a blood sample was taken to provide a DNA record. The record will prove useful if the eagle is found dead, whether of natural causes, or unnatural poison.

The latest arrivals were part of the second wave of an ambitious reintroduction programme led by Scottish National Heritage (SNH), and supported by the Royal Society for the Protection of Birds (RSPB). Between 1975 and 1985, a total of eighty-two sea eagles were freed on Rum, but failed to establish a viable population. Twelve of them were recovered dead, including one killed by a shotgun. The second initiative was launched in 1993, and around ten birds have been released each year since. Yet by the summer of 1997, less than ten pairs had bred successfully, and the population will only be judged safe when more than twenty pairs are producing young.

The fact that these great birds are once again being seen by fishermen, farmers and tourists in the west Highlands and over the Hebrides is no matter of eco-whimsy. The

Government is charged, by EU and international legislation, to enhance habitats and – where possible and practicable – to reintroduce extinct species. Another raptor, the red kite, has been successfully, and non-controversially, re-established in Scotland and England. Ospreys, also persecuted to extinction, have returned of their own volition, and are producing up to 100 young each year.

But the sea eagle has been less fortunate in its welcome in the hills and glens. Depending on your point of view, the restoration of the super-predator to its old haunts, on a seaboard over which it once held dominion, is either a spectacular addition to the fauna of Britain, and part of an urgent Europe-wide protection programme for an enigmatic species; or, it is meddling, misguided conservationism, which fails to consider the interests of the people who make a living from the land.

The latter point of view is held by a very small minority. The sea eagle was hunted out of Scotland by gamekeepers and shepherds who shot it because it preyed on lambs. But the bird's diet of choice is fish, sea birds and hares. It will, in spectacular fashion, harry the eider duck and the shag to exhaustion, forcing them to dive repeatedly in a futile bid to escape. However, like most birds of prey, it is also opportunistic. It will feed on carrion where it can find it, and it will take lambs, alive or dead, if they present the best option. So, predictably, the old enmities have surfaced again.

On one large Scottish island, a 'rogue' pair of sea eagles took at least seventy lambs over a period of five years from 1990. To the farmers involved, its return was unjustifiable. They were compensated for their losses, but would rather have sold their lambs for fattening than have them eaten by birds of prey for the somewhat intangible greater good.

The problem is instructive in one important respect: lambs would not be a major part of the bird's diet if other food was available. The implication is – and this should be familiar territory to scholars of the environment – that over parts of the Highlands, the dominance of mono-cultural, extractive regimes of hill sheep farming and grouse moor means that predators at the top of the food chain have little choice but to dine on lamb or game bird more often than is natural.

I do not believe, as I have already indicated, in the growth of native animal populations at all costs. But the sea eagle hardly falls into that category, and its return is opposed by neither the Scottish Landowners' Federation, nor the National Farmers' Union. On balance, there is no doubt that the eagle with the sunlit eye should be welcomed home. It is also right that the Government should be charged to consider the reintroduction of species which were removed, or marginalised, by man.

In many cases, extinctions cannot be overcome because the beasts involved, particularly the predators, have lost their habitats and their prey species. But where reintroductions can be effective, it may make sense from the economic and environmental points of view. Wildlife tourism, on its own, or as part of a holiday package, is said by the Scottish Tourist Board to generate £60m a year. It is one of the biggest growth sectors in the market – even in the 'man-made desert' which has lost its most compelling characters (animal and human?).

Our bestiary is depleted, our native habitats altered and diminished, and any balance in wildlife populations largely gone or threatened. In such a situation, the species which remain should be treasured all the more and, if possible, encouraged. The sea eagle is important in its own right because it was our biggest bird of prey and because it was us, through ill-advised persecution, who drove it to extinction. If we can bring it back, it will prove that times have changed and that we are thinking more carefully about our relationship with nature. The same bird is threatened throughout Europe, and should be counted as one of the more remarkable of Scotland's 90 000 species – most of which have to be seen through a microscope!

So how far can we take the argument? Well, there are people who would like to see the wolf and the lynx roam the high tops once again. There is even a brown bear society in Jersey which wants the big hunter back in Scotland. The principle is laudable, but the difficulties for such ambitious reintroductions are many and varied. It seems unlikely that we have the right habitat and countryside management policies at present for the return of the wolf, for example, although it has sneaked back into Germany across the River Odra from Poland, to begin breeding in the wild again for the first time in 150 years.

A more likely candidate for Scotland is the European beaver, a large herbivore which was once common in our fresh waters, but was hunted to extinction for its pelt and scent gland. In the tenth century, its skin fetched 120 pennies – compared to 18 pennies for an otter pelt – and by 1188 it could only be found in the River Dee in the north east of Scotland. It eats willow, aspen, poplar, grasses and the twigs of herbaceous plants.

Concerned landowners do not object to its diet but to its dam-building habits which they believe could damage the upper reaches of salmon rivers, and threaten spawning grounds. The evidence from France, where it has been successfully reintroduced, would suggest otherwise. SNH is considering the beaver's return, and will go through a comprehensive consultation process before a final decision is made.

Striking a balance is important. We cannot do too much, too soon, and must take the majority of the local populace on the slow ride to a healthier, more diverse, Scotland. For my own part, I have been thrilled since childhood, when my father pointed out golden eagles above Ballachulish, or showed me herds of red deer in the winter snows of Glencoe, by Scotland's wildlife. I have also been lucky in more recent times to see the white-tailed sea eagle fly again in the west Highlands. Its nickname, the flying door, springs readily to mind when it appears above the mountains of Wester Ross.

Ideally, its return should symbolise the health and vitality of its rediscovered environment. Unfortunately, we are not yet at that stage.

Scottish Wildlife in Perspective

The word biodiversity, a shorthand for biological diversity, was coined as recently as 1985 by Walter Rosen, an American scientist, at a conference in Washington DC. It is now enshrined in international treaties and government planning documents all over the world. Biodiversity in Scotland does not compare with the teeming life of the equatorial zones. But biodiversity everywhere, including Scotland, is the essence of life and is vital to our survival. Worldwide, there are 30000 species of edible plants, but just twenty of them provide 90 per cent of our food. Of those, wheat, maize and rice account for half of all consumption. In recorded history, no more than 7000 plants are thought to have been grown or collected as food.

Globally, the rate of species extinction is running at around 300 a day. Many of these may be microscopic creatures or small plants, but big animals, ranging from the sea eagle to the panda, are also under threat.

In his book *The Diversity of Life*, the biologist Edward O. Wilson illustrated the potential of the species we are losing with the story of a remarkable discovery in the 1970s by a Mexican college student. He found the maize plant *Zea diploperennis*, a wild relative of corn, on 24 acres near Guadalajara at the point when it was one week away from extinction by fire and machete. Close examination of the maize revealed that not only was it resistant to disease, it was also unique because it possessed perennial growth. If its genes were transferred to domestic corn, it would boost world production in dramatic fashion.

When I think of biodiversity, I recall a passage early in Wilson's book in which he describes a thunder storm in a Brazilian rainforest.

> Somewhere close I knew spear-nosed bats flew through the tree crowns in search of fruit, palm vipers coiled in ambush in the roots of orchids, jaguars walked the river's edge; around them 800 species of trees stood, more than are native to all of North America; and a thousand species of butterflies, six per cent of the entire world fauna, waited for dawn.

About the orchids of that place we knew very little. About flies and beetles almost nothing, fungi nothing, most kinds of organisms nothing. Five thousand kinds of bacteria might be found in a pinch of soil, and about them we knew absolutely nothing. This was wilderness in the 16th century sense, as it must have formed in the minds of the Portuguese explorers, its interior still largely unexplored and filled with strange, myth-engendering plants and animals.

From such a place the pious naturalist would send long respectful letters to royal patrons about the wonders of the new world as testament to the glory of God. And I thought: there is still time to see this land in such a manner.

The time has gone to see Scotland in this manner. But Scotland too has its own unique, and still fascinating, assemblages of plants and animals.

The Resource

.

Mammals

Since the loss of the land bridge between Britain and Europe, at the end of the last Ice Age, eighteen mammals have been added to the Scottish bestiary deliberately or accidentally – but rarely happily – and twelve have become extinct. We have lost some wildlife stars.

The wild horse, the reindeer and the elk went, as the climate changed, around 10 000 years ago. The lynx, one of the great carnivores of old Caledonia, went as long ago as 4000 BC. The formidable Caledonian brown bear was admired for its size and good looks and was exported to the Coliseum in Rome where criminals would be tied naked to the cross and left to its mercy. It hung on until 500 AD.

The wolf went as recently as 1743, when the last one was said to have been shot by a hunter named MacQueen by the River Findhorn in Morayshire. A rival candidate for the melancholy title of 'the last wolf' was killed by Cameron of Locheil at Killiecrankie in 1680. Whichever claim is correct, the wolf's departure was inevitable. It was persecuted all over the country by professional hunters who made large sums of money from the rewards paid for wolf pelts. There was a professional wolf hunter in Stirling as early as 1283, and the last healthy wolf population existed in the time of Mary Queen of Scots.

All the extinctions can be explained in terms of climate change, loss of habitat, the expanding human population and persecution.

What of the introductions? The feral goat, which contributes with singular efficiency to the destruction of oak and pine forests in many parts of Scotland – on Loch Lomond and in Wester Ross – was brought here 8000 years ago by early farmers; the house mouse arrived around 4800 years before the present; the rabbit, perhaps the least welcome of all introductions, appeared in the late twelfth century; the common rat in the seventeenth century; the grey squirrel in the late nineteenth century; the predatory and hugely destructive American mink – in the form of escapees from mink farms – just sixty years ago. And still it goes on. It is thought that the tiny muntjac deer – more bad news for native vegetation – walked across the Border into Scotland as recently as 1994. One conservationist remarked in the summer of 1997 that muntjac deer should be 'shot on sight' and was reprimanded by animal welfare groups for saying so. But he was right. Muntjac deer should be shot on sight.

Today, when we think of wildlife in Scotland, we continue to concentrate on the more charismatic creatures which inhabit our moors, forests and farmland. Everyman's instant list of wild mammals would include the red deer, the red squirrel, the fox, the badger and the otter. But our varied habitats are also home to the wildcat and the pine marten, both with populations of around 3500, the sika deer, and no fewer than eight species of bat.

In total, there are 62 mammals in Scotland, with 48 breeding on land and 14 found in coastal waters, of which nine have been seen with young. The abundance and relative health of the species varies enormously. How many people know that Scotland's most populous mammal is the field vole? There are 41 million of them, while the red-necked wallaby population (an introduction limited to one island in Loch Lomond) stood at 28 individuals, at the last count.

Small areas of the country can lay claim to entire British populations, such as the Orkney vole, the reindeer (reintroduced in the Cairngorms) and the wildcat. In general terms, although we are not heading immediately for another mammalian extinction, the status of many of our wild animals is not as it should be. Of the 48 land mammals, 12 are known to be declining, and the status of others is unknown. The mammals which are increasing in number are those most associated with habitat damage, like the red deer, the sika deer – which is inter-breeding with the red deer and destroying its genetic purity in many mainland areas – the grey seal, and the grey squirrel, which is squeezing the native red squirrel.

This simple summary should be enough to reinforce the point that our 'wild' environment is greatly influenced by man, and therefore has to be positively managed to benefit, or control, mammalian species.

Persecution is no longer the main cause of decline, although gamekeepers, often with the approval of the landowner, illegally kill birds of prey, badgers and pine martens. Wildlife is also threatened by road traffic, pesticide poisoning, the acidification of fresh waters by air pollution, the loss of habitats to development, genetic mixing and climate change.

Insectivores
There are five native insectivores – the hedgehog, mole, common shrew, pygmy shrew and water shrew – whose numbers fluctuate dramatically from year to year. They suffer from habitat loss, encouraged in the past by subsidy payments to farmers to increase production and the area of productive land, and from intensive farming which requires the use of pesticides.

There is some hope, however, that the modern trend in agricultural subsidies, which are moving towards the goal of countryside stewardship, will lead to an increase in numbers. For example, grants for the set-aside of agricultural land, and for the expansion and creation of farm woodland, will help some species. The five insectivore populations are relatively high, but we do not know whether numbers are increasing or decreasing.

Bats

Scotland has eight native species, all of them protected under the 1994 European Bat Agreement and the Wildlife and Countryside Act (WCA). Counting bats is problematic unless they roost in houses. Consequently, only one species, the natterer's bat, is known to have a stable population, and one other, the whiskered bat, is known to be in decline.

Population stability is determined mainly by the abundance of insect prey, which can be influenced by habitat and land use, altitude and latitude. The commonest bat is the pipistrelle, with a population of around 550 000 – more than five times the number of the other seven added together. One of the main threats to bats has been the use of toxic chemicals in the treatment of dry rot and woodworm in attic roost sites. In the past decade, the identification of the problem chemicals has helped reduce the threat.

Lagomorphs (Rabbits and Hares)

The mountain hare, which has an attractive white coat in winter, is the only native lagomorph. Almost the entire UK population – around 350 000 – is found north of the Border, where numbers are declining due to habitat loss, competition and persecution. The Government is now required to safeguard the species under the EU legislation, but the hare is already restricted to the high tops as a result of pressure lower down the hill from the brown hare, which was introduced to Britain 2000 years ago.

Rabbits – all eight million of them, despite the collapse in the population caused by the deliberate introduction of myxomatosis in 1953 – illustrate the problems that can be caused by non-native species.

The disease killed 99 per cent of Britain's 100 million rabbits, and the serious infestations which affected 55 per cent of Scottish farms dropped to 0.4 per cent of farms. By

1991, the figure had climbed to over 16 per cent, and is still rising.

First introduced around 800 years ago, they were abundant around Edinburgh in the 1500s, but did not become established throughout the country until the early nineteenth century, when efforts were made to establish colonies for food and sport. They caused an estimated £12m worth of damage to agriculture in 1994.

Rodents (Gnawing Animals)
Scotland has eleven rodents, of which five are native. The red squirrel is the only one protected under the WCA, and is thought to be in decline in Scotland, following the trend in England and Wales, where the advance of the non-native grey squirrel has proved disastrous. The greys were introduced to southern England 200 years ago, and have outcompeted the red over large areas. As a result, Scotland has the dubious honour of holding 75 per cent of the red population – 121 000 animals – principally in native pine woods, and especially in Deeside and Speyside. There is anecdotal evidence that the grey is advancing northwards – it has certainly reached the Great Glen – and there are serious doubts about the future of the red squirrel.

Another native rodent, the water vole, is not protected by law, but appears to be suffering a dramatic decline in numbers, mainly due to habitat destruction and predation by the most disastrous introduction to Scotland after the rabbit – the voracious American mink, once farmed for its fur and now thriving almost everywhere, with a population of around 52 000.

Scotland's two most populous mammals are rodents, with 41 million field voles outnumbering every land and sea mammal added together, and no fewer than 15 million wood mice. Another unwelcome addition is the common rat which causes health hazards and serious problems for bird populations. On the island of Handa, off the Sutherland coast, rats are preventing the return of the guillemot and limiting the spread of the puffin, and may be preventing the spread of fulmars.

Carnivores

Carnivores are at the top of the food chain, and are among Britain's rarest animals. The otter, wildcat and pine marten – which are rarely seen – are protected by law.

The otter, plentiful in Skye and the Northern Isles, is also protected under the EU species directive which demands that areas of feeding and breeding habitat are identified and protected. Overall, while reintroductions are practised in England, the Scottish population seems to be stable or even increasing, although disturbance and pollution have driven it out of some areas. It is a relatively short-lived animal, and some keen observers are concerned that an accumulation of pollution at sea and in rivers – of mercury, for example – could lead to a quite rapid collapse in numbers. There are thought to be just 6600 otters, but the figure is accurate only to plus or minus 50 per cent. Habitat, disease and road-loss are the major known causes of fatalities, yet the otter is quite hardy in the right circumstances. With very low numbers in England, and the loss of the otter in Belgium in 1986 and the Netherlands in 1989, the healthy Scottish population is growing in importance, and is a useful confirmation of the relative cleanliness of our fresh waters. The Swiss, for example, have given up trying to reintroduce otters because their rivers are too polluted. Otters were said to be under threat during the building of the Skye Bridge, but the animal is highly adaptable, and in some areas has taken to using the rock armour used to support roads for its holts.

The status of the wildcat is currently under investigation by scientists concerned that the real thing may no longer exist due to decades of inter-breeding with the feral cat. Scottish wildcats were isolated at the last Ice Age, and enjoyed 8000 years of race purity before the Romans arrived with domestic cats. The feral cat is now recognised as the same species, and scientists have, as yet, been unable to distinguish between true wildcats and feral cats of domestic origin. In the current situation it is almost impossible to guarantee the protection of Scotland's estimated 3500 wildcats while there are 130 000 feral cats. The real thing is protected by law, while the feral cat is not. When SNH scientists examined 400 live and dead cats, genetic testing only established groups by geographic area, and did not distinguish between wild and feral animals. The point was proved in a court case at Stonehaven in 1990, when a game-keeper was accused of killing three of the rare animals, but was not convicted because no-one could say 'beyond reasonable doubt' what they were.

The situation is more promising for the pine marten although its population is low, at 3500, and it is extinct south of the Border. It has been persecuted for centuries – and still is today – for its killing instinct around farm livestock, particularly poultry. Before it was protected by legislation, it was pushed by persecution into the north-west of Scotland. With its newfound status, it has been spreading south again. There has been one unconfirmed sighting in Killiecrankie, and there are reliable reports of a marten fan releasing a lorryload of the attractive animals in the Borders. In response to the concerns of land managers, SNH has been working on pine marten-proof pheasant pens. Thankfully, persecution today is much less strenuous today than it was in the 1800s, when hundreds of martens would be killed on a single estate in a single year – a fact which helps to explain the size of the current population.

Polecats, once hunted to extinction, were reintroduced in a private initiative in the 1980s but their current status is unknown. They have their supporters, who believe they would help to control the rabbit population in some areas, and their detractors – the successors to the landowners and farmers who got rid of them in the first place.

Another charismatic, but rarely-seen, nocturnal native is the badger. Its population is thought to stand at around 25 000, representing just 10 per cent of the UK population. It is doing pretty well all over the country, but volunteer badger groups have been set up – in Fife, for example – in an attempt to protect local populations from disturbance. There are still regular reports from central and southern Scotland of men digging up badgers to bait them in fights against dogs. Badger baiting is reprehensible, and one of the worst forms of animal persecution still practised in Scotland.

The distinctive animal is also unpopular for digging up root crops.

The American mink is another thing altogether. Imported for its fur in 1929, it was breeding in the wild by the 1950s after dozens of accidental escapes and deliberate releases. Since then, it has been working its way into new territory, decimating ground-nesting bird populations on the way. The release of animals from a farm which failed on the Western Isles has also allowed the mink to threaten breeding birds on Lewis and Harris, and, sad to say, it has been seen on North Uist – having somehow crossed the North Channel – where it threatens one of Britain's rarest birds, the corncrake. There are a few Government-sponsored trapping programmes, but the animal is notoriously difficult to catch and kill in significant numbers. Eradicating the mink is thought to be impossible, and the aim is to control it in key wildlife areas.

Another carnivore which is not endangered – and which does not get much protection – is the fox. There are around 1000 foxes killed between Lochaber and Perthshire each year and the Scottish Office funds fox control groups, including three packs of hounds at Aberfeldy, Tomatin and Strontian, to help control the animals and prevent them taking lambs. There is no doubt that foxes take lambs and game birds.

Ungulates (Hoofed Animals)
There are seven ungulates breeding in Scotland, of which only the roe and red deer are native. The red deer is Scotland's largest land mammal and herbivore and its numbers have increase dramatically in recent decades. The current population, which is guessed to be 350 000 – and may be higher because of the difficulty of counting deer in woodlands – has caused serious habitat degradation in many parts of the country. Over large swathes of the Highlands, particularly in the Cairngorms, the absence of natural regeneration in native and semi-natural woodland can be attributed to the combined effects of deer and sheep.

The signs are obvious everywhere. On either side of the

A9 the bare moorland stretches away over smooth horizons, and the view is much the same from the A82 across the Moor of Rannoch and through Glencoe. Trees such as rowan and birch are restricted to the few spots on the landscape which are out of reach of grazing mouths, and so sprout from the top of erratics and out of cliff faces and dark defiles.

Red deer are a vital part of native Caledonian woodland, and are highly valued for sport, but it is almost universally accepted that in many parts of the country their numbers are out of kilter with the natural habitat. The unacceptably high numbers are maintained by winter feeding, and by the failure of sporting estates to cull enough animals, particularly hinds, each autumn and winter. Scotland's deer are remarkable, however, for having adapted to the open hills and, in the process, have become the world's smallest and lightest red deer.

Meanwhile, another incomer, the non-native sika deer, threatens the genetic integrity of the red deer through hybridisation. Studies in Kintyre have established that a high proportion of deer in the area are hybrids, although interbreeding is not always obvious to the naked eye.

Scotland also has a large and increasing population of roe deer, which are found in upland and lowland areas.

Another animal playing a part in the stagnation and decline of oak and pine woodlands is the feral goat, whose numbers are thought to have stood at around 2600 since the 1960s. Although that figure does not seem high, goats are found in rugged terrain associated with remnants of native woodland, and can play an important role in habitat damage.

Seals

Grey seals and common seals (the less common of the two) are protected by European law, although they can be shot if they are interfering with fishing gear of salmon farms. Their protected status is a subject of controversy with fishermen, anglers and the landowners who have salmon fishing rights. Fishermen claim grey seals eat two tonnes of fish each a year, which adds up to 200 000 tonnes per annum, or more than the total catch of the Scottish white fish fleet. They also believe that seals have played a significant role in the drop of haddock and whiting stocks over the past two decades, and inshore fishermen say seals have learned how to break into their creels and steal the bait.

While the common population is stable at around 33 000 animals, the population of grey seals has been increasing at an average rate of at least 6 per cent per annum for the past ten years and is thought to stand at around 100 000 animals. The last seal cull in the UK was held in 1978 and recent demands from fishermen for another cull have been made without conclusive proof that the mammals are destroying commercial fish stocks. Conservationists also point out that around 75 per cent of the world's population swims in Scottish waters.

However, it is also true that the continuing protection afforded to the seal is based to a considerable extent on the judgement that another cull, while it might not damage the overall population, would be unacceptable to the general public. In countries such as Norway, however, the annual cull does not threaten the viability of local populations.

Cetaceans (Whales, Porpoises, Dolphins)

As many as nine species may breed in British waters, including dolphins, porpoises and the minke whale, which survives in large numbers in the North Atlantic and is the subject of a controversial small fishery in Arctic Norway.

The minke excepted, the population status of cetaceans is not well known, and work is continuing to estimate local populations. According to SNH, it may be possible to utilise genetic technology to assess the relative health of whale populations by determining levels of inbreeding. At present, it is true to say that the population trends among the large whales visiting Scottish waters – the humpback, the fin whale, the minke, the sperm whale and the northern bottle-nose whale – are unknown. Only the minke is commonly sighted, although there have been a series of strandings of sperm whales on the east coast in recent years. Sperm whales should migrate south to equatorial waters off the west of Scotland, and their appearance in the North Sea has alarmed conservationists, who have suggested – without hard evidence – that oil development west of Shetland may have driven the great mammals off their normal route and into the enclosed North Sea, where squid, their chosen prey, are in short supply.

Among the small whales, the harbour porpoise, common dolphin, bottlenose dolphin, white-sided dolphin and long finned pilot whale are frequently sighted, and the white beaked dolphin, the killer whale and risso's dolphin are common. Once again, however, the population trends are not known.

The common threats to cetaceans include pollution, and disturbance and injury from propellors. Carcasses found washed ashore suggest that incidental catches from the fishing fleet, and what is known as MTS (multiple trauma syndrome) in porpoises, especially in the Moray Firth, are major causes of mortality. Porpoises in the Moray Firth have also been killed in attacks by much larger dolphins.

Birds

Scotland is enormously important for birdlife, and many populations are increasing in its relatively clean fresh waters, estuaries and coasts.

Between 300 and 325 bird species are recorded annually, including 200 residents or regular summer, passage or winter visitors. Birds are more readily counted than many animals, and so there is reliable census information for a large number of species.

Scotland's bird life is rich because it is at the junction of two major 'flyways', one from the high Arctic Canadian islands, crossing Greenland and Iceland, bringing the geese which winter in Scotland, and the other from the east, from northern Russia across Scandinavia, bringing waxwings and dotterel.

The most important species for conservation include the white-tailed sea eagle, the corncrake, whose numbers have collapsed due to intensive farming techniques and the loss of habitat, and the red kite, another raptor which has been successfully reintroduced to Scotland.

There are, however, a total of 117 species in the so-called *Red Data Book* which lists endangered species – including the capercaillie, the black throated diver, the whimbrel and the golden eagle. Just over ninety of the species listed occur regularly in Scotland.

Divers and Grebes
These birds thrive on Scotland's typical oligotrophic (nutrient-deficient) lochs in the north and west. There are just 150 breeding pairs of black-throated divers in Britain (all of them in Scotland), 1350 red throated divers, and only 74 pairs of Slavonian grebes.

The trend in all three species is downwards, for reasons including acidification, predation of eggs and, in the case of red-throated divers, the availability of food at sea. Black-throated divers are the subject of an ambitious conservation programme in which artificial nest platforms have been built to encourage their return. In some cases, these are placed in sites intended to lure the birds away from lochs affected by hydro schemes, where nests can be flooded by changing water levels.

Wildfowl
Scotland is important for breeding and wintering wildfowl. Among the former there are sixteen key species, most of which are increasing in number, with the exception of the common scoter, which is down to 100 breeding pairs, and declining. The variability of the populations is illustrated by the recent decline in scaup and goldeneye. Wintering wildfowl include the Greenland barnacle geese on Islay and the Solway, and the Greenland white-fronted geese on Islay. The protection of these birds in the Arctic and in Scotland means numbers are increasing, along with the compensation payments made to the Solway farmers and Islay crofters whose grass is eaten and land trampled.

For many Scots, the arrival of cackling skeins of geese from the far north is a potent symbol, and an attractive reminder, of impending winter. The most numerous are the pink footed geese, numbering around 200 000.

Seabirds

Scotland is also internationally important for the four million seabirds which breed on the sea cliffs of Orkney, Shetland and the north-east. Millions more nest on St Kilda, and on other small island groups nearer the mainland. Scotland boasts 49 per cent of the world's northern gannets and the numbers of most species are thought, at present, to be stable or increasing. The upward trend was helped by the end of the widespread persecution of seabirds. Where there have been significant declines they have been associated, according to conservationists, with industrial fishing for sand eels, a basic food for many seabirds. For example, numbers of roseate and arctic terns have suffered poor breeding seasons in Orkney and Shetland.

Raptors

There are ten 'priority' species in Scotland, where, unfortunately, deliberate killing of birds of prey continues on some sporting estates. One raptor which suffers from the illegal methods of some gamekeepers and landowners is the hen harrier. There are around 630 pairs in Scotland, and the future is a testament to the Victorian attitudes which still prevail among many absentee lairds. Left to its own devices, the hen harrier will create moorland colonies, but no colony is known in mainland Scotland. The harrier and the peregrine falcon – population 1100 – are blamed by many landowners for poor grouse numbers on sporting estates. There is no doubt they do prey on grouse, but the likelihood is that they are only one part of the equation in the cyclical nature of grouse populations. They are a particular threat to grouse on grassy moors where the birds have limited cover.

There are eminent ecologists in Scotland who believe it is possible to maintain high grouse numbers without killing birds of prey, simply through better, and more vigorous, land management. Birds of prey are easy targets.

The peregrine is a target, but its numbers have increased significantly in the past three decades following a population crash caused by the use of pesticides such as DDT on farm fields. The chemicals are now banned.

Meanwhile, the golden eagle, a talismanic bird for Scotland, has a disappointing breeding population of around 400 pairs. The figure appears to be stable, but the eagle does not breed over vast ranges. In the Monadhliaths the territory appears to be ideal. But more than ten eagles have been found dead in that mountain area since 1973, including one in an illegal gin trap and five confirmed poisoned victims. In the same period, in the same area, poisoned buzzards, sparrowhawks and hen harriers have been found, and peregrine falcon nests or chicks have disappeared. Golden eagles are found poisoned and shot almost every year in Scotland, and it is assumed they have been shot and persecuted out of certain areas. Who is responsible? Out of twenty-five court cases dealing with illegal killing since 1981, twenty-four have involved gamekeepers.

Waders

The British populations of eight species of wader are confined to Scotland. These include the turnstone, greenshank and the rare red-necked phalarope. Afforestation of hill and moorland areas has affected wader numbers, as has the loss of lowland wet grassland. The reclamation of coastal land affects passage and winter waders using Scotland's estuaries and open coasts.

Passerines

Of the twenty-five 'priority' species, the breeding populations of thirteen are found only in Scotland. The most endangered member of the group is the corncrake, whose numbers have been in rapid decline due to changing farming methods and the earlier cutting of silage in July, which results in the slaughter of adult birds and chicks. The corncrake is now hanging on in the Hebrides, and there are small increases in some areas – Coll, for example – due to conservation programmes involving the RSPB and SNH, which offer payments to farmers for corncrake-friendly farming. This means cutting fields after 1 August, which allows the chicks to fledge. Farmers are also encouraged to leave rough field edges and corners to support the birds, and to cut their fields from inside to out, giving the birds a better chance to move to safety. Corncrakes depend on good farm management; they need rough margins, nettles, irises and cow parsley for nest sites and protection. If the land is abandoned and the fields become rank and infested with rushes, they will die out. It always seems remarkable to me that these idiosyncratic visitors undertake a huge annual migration from South Africa, only to spend six months skulking in what is left of the Hebridean undergrowth! Overall, the numbers are slightly up in recent years with around 600 singing males in 1996.

The capercaillie, Britain's biggest game bird and a startling sight to behold in native pinewoods, became extinct in the late eighteenth century, possibly through the felling of native forests. It was reintroduced in Perthshire in 1837, but is now in decline again. The 1992–4 census showed a population of around 2200, possibly one tenth of the number surviving in the 1970s. The birds are mainly concentrated in Perthshire, Strathspey and Deeside. They appear to favour old trees in open forest and nest on the ground among heather and blaeberry. The control of predators is therefore of great importance to their survival. However, recent research suggested that up to one third of capercaillie deaths are caused by collisions with deer fencing.

The grey partridge is also in rapid decline, as are the corn bunting, tree sparrow and song thrush. A report commissioned by the Government in 1997 suggested that twelve species of farmland birds were in decline because of pesticides said to be 'drenching the countryside in poison'. These included the turtle dove, skylark, reed bunting and swallow. There are also concerns over the status of garden birds such as the blackbird.

Finally, Scotland's only endemic bird – the only one found here and nowhere else – is the Scottish crossbill, *Loxia scotica*, which inhabits the old Caledonian pine forest and extracts seeds from pine cones. It specialises in Scots pine, while the European crossbill – an annual visitor from Scandinavia – will eat seeds from softer spruce cones.

Amphibians and Reptiles

Scotland has nine of the twelve species native to Britain, including three non-native species – the alpine newt, sand lizard and grass snake. Coastal waters are regularly visited by marine turtles, particularly leathery turtles, which are rarely seen but occasionally found in fishermen's nets.

There are just three native terrestrial reptiles which are widespread and common in semi-natural habitats – the slow worm, vivivperous lizard and adder. A recent survey of adders established that their greatest concentrations are in south-west Scotland and the central Highlands, and that there are areas, in Argyll for example, where they should be found but are unaccountably missing. It also suggested a decline in numbers on farm land.

Frogs (Scotland has just one species), common toads, smooth newts and palmate newts are the most common amphibians (living in water and on land) and are found throughout the country, with the exception of some islands. The natterjack toad and the great crested newt are threatened in Britain and Europe, with just forty natterjack colonies in the UK, four of which are on the Scottish shores of the Solway Firth, where they are at the northern limit of their European range. Great crested newts are commoner than natterjack toads, but a national survey recently found the amphibian in only 56 of 1500 Scottish ponds, mainly in the south-east and the coastal fringe of Dumfries and Galloway.

Fish

There are more than 160 species of fish in Scotland's inland and coastal waters. Some fish, including the arctic charr and powan, are only found in fresh waters, while the flounder is found in estuaries and will tolerate a wide range of 'saltiness'. Migratory fish, like the salmon and sea trout, traverse all three habitats.

Our freshwater fauna is substantially impoverished compared with the fish populations elsewhere in Britain and Europe. However, forty-two out of the fifty-five species found in Britain occur north of the Border, including species which are little understood.

The original colonisers after the last Ice Age used the land bridge which survived between England and the continent for around 3000 years, and include the familiar brown trout, Atlantic salmon and three-spined stickleback, and the less well known powan, sparling, sea bass, thick-lipped mullet and twaite shad. By the end of the eighteenth century, the pike, minnow, roach, stone loach and perch had been introduced.

The new arrivals began in the late nineteenth century with brook charr, grayling, tench and bream, and by 1970 we also had the rainbow trout, pink salmon, goldfish, common carp, dace, bullhead and orfe. Scientists believe these may have been transported by humans or could have been carried as eggs on the feet of wildfowl, or in water spouts. Strange but true. The most recent introduction, and one of the most damaging, was the ruffe, one of many species used as live bait and discarded in lochs by visiting coarse fishermen from England. The ruffe, a small perch-like fish, has exploded in numbers in Loch Lomond, and threatens to wipe out the native powan, one of Scotland's rarest fish, which occurs only in Loch Lomond and Loch Eck.

Many introductions were the bright ideas of landowners who wanted something exotic in their ponds and hill lochs. Fish which failed to become established include the large-mouth and smallmouth bass, brought to Loch Ba by the Duke of Argyll in 1881, and the cutthroat trout which was released in Shetland. One landowner in Fife tried, in the 1920s, to introduce American lake charr, dolly varden, the Danube catfish and the quaintly named pumpkinseed to his ponds between 1920 and 1930. He failed. But the incomers brought with them disease, competition and predation.

In general terms, freshwater fish, including the ubiquitous brown trout, are undervalued and under threat. Key threats include industrial and domestic effluent, acid deposition, river obstructions, drainage, fish farming, angling and global warming.

Meanwhile, despite fluctuating numbers, our most famous game fish, salmon and sea trout, continue to be of major economic benefit to rural areas. Salmon angling alone is said to be worth £50m a year, but is affected by habitat damage, illegal fishing at sea, acidification, genetic contamination from escaped farmed fish and global warming. Catches of salmon by rod and net have varied greatly since the advent of records in the early 1950s. Until the 1960s, the annual catch figure was around 400 000 fish, rising to between 480 000 between 1962 and 1973. A sharp decline to a figure of around 300 000 has occurred since 1974. The drop coincided with the advent of monfilament drift net fishing on salmon migration routes in the North Atlantic and off the Northumberland coast. However, many commercial salmon fishing operations, both at sea and in Scottish river estuaries, have been bought out by an international alliance of angling interests, and soon all the salmon caught in Scotland may be taken on rod and line.

Unfortunately, there is still no overall administrative framework for the proper managment of fish populations and fisheries. All the scientific research in recent years has favoured the Atlantic salmon because of its value as a game fish.

In the meantime, Scotland's freshwater fish populations have been so altered over the centuries that only a small number of lochs in the remoter parts of the country are likely to have truly native stocks. Game fishermen are still introducing American rainbow trout to brown trout lochs, while many lochs have been stocked with trout taken from other parts of the country, of a different genetic stock. And we have no idea of the status of charr populations, which are generally undervalued as a wildlife and tourism asset, and as a food source.

In the current situation, it does not take a leap of the imagination to suggest that legislation to control the introduction of new species and the transfer of fish across natural watershed boundaries is required.

Around the coast, estuarine fish change with the seasons and the years. The main threats they face are land reclamation, the construction of barrages and over-exploitation. The Firth of Forth, for example, has forty-three species including seasonal residents such as cod, whiting, herring and sprat. An improvement in water quality in our estuaries has resulted in the return of salmon to the Clyde and sparling (a fish which smells like cucumber) to the Forth. There are currently thirty-four fish species recorded in the Inner Clyde, where there were none known in 1850.

I have chosen to concentrate on our freshwater fish, but in the deeper waters around Scotland it is possible to find cold-water species like the halibut, and warm-water species such as the sun fish. The main commercial marine species – the cod and herring family – are best understood but most at risk from exploitation, while a review of rare British fish has also identified the basking shark and the common skate as species at risk because of their low reproductive rates.

Invertebrates

There are more species of invertebrate than anything else in Scotland, with around 14 000 different insects, 8000 marine invertebrates and numerous nematodes or flatworms.

Terrestrial Invertebrates

Around 55 per cent of the British insect species are found in Scotland, including many unique species found on distinctive mountain and woodland habitats. Of the 14 000 species of insect breeding in Scotland, 1300 are not found elsewhere in the UK. The Scottish fauna has a strong affinity with Scandinavian invertebrates, but very few are endemic, probably because of the last Ice Age.

Most butterfly populations are thought to be constant or increasing in numbers, with some – the speckled wood, orange tip and ringlet – expanding. The small blue is thought to be stable, and the chequered skipper has its stronghold in colonies in the west of Scotland, following its extinction in England. Another extinction in England, the New Forest burnet moth – which disappeared from the New Forest in the 1920s – was discovered in the 1960s at one site in the west of Scotland. By the 1990s, its habitat was declining due to over-grazing by sheep, and fencing has been used to stabilise the population.

Invertebrates are particularly useful as indicators of climate change, and work is underway at present on the distribution at altitude of the micro-moth, which adjusts its range according to temperatures. At lower levels, the developing conservation management of native woodlands is leading to the spread of many Scottish insect species, and invertebrate populations now have to be considered by planners.

Freshwater Invertebrates
There is limited information on Scotland's freshwater invertebrates, although our knowledge is slightly better for river fauna such as water beetles, dragonflys, damselflies and the freshwater pearl mussel. Twenty species of water beetle, ten of caddis fly and two of dragonfly are on the *Red Data Book* list for endangered species, and the freshwater pearl mussel, threatened by pollution and over-fishing, is also protected. Of the 69 Scottish rivers which once had pearl mussels, 26 now have none, nine have a handful, 31 have low numbers, and just three have good populations. It is illegal to kill or injure a mussel, but not to fish for pearls.

There are several non-native freshwater invertebrates in Scotland, including the New Zealand flatworm, introduced by the timber and aquarist trades with potentially serious effects on native fauna, particularly the earthworm.

Marine Invertebrates
Although these species have been present for thousands of years they are not well known, but appear in a startling array of colours and shapes, and range from the barnacle to the octopus and jellyfish.

Current concerns include the commercial harvesting of cockles and razor shells. The seabed and its fauna are also disturbed by commercial trawling which can damage non-mobile species like sea pens. The feed and chemicals from salmon farms can also damage the sea bed flora and fauna around fish cages. Scotland's marine invertebrates need to be better understood.

Flowering Plants and Ferns

There are 1600 species growing wild in Scotland, ranging from floating water plants to large trees. Large numbers of species have been introduced accidentally or deliberately. The New Zealand willow herb, for example, can be found on the remotest mountains and islands, but with no obvious effect on native species. Others, in particular the ubiquitous rhododendron (*Rhododendron ponticum*), have greatly reduced the diversity of native species and the habitat available to them.

The number of species in Britain is small by world standards, and insignificant compared to the areas of endemism where, for example, more species of plant are found in a single Ecuadorian volcano than in all of North America. Scotland also suffers because plants had to re-colonise after the recent Ice Age.

The small number of endemic plants include the Scottish primrose, which is confined to a narrow coastal strip in the far north and Orkney, the Scottish small-reed, found only in Caithness, and two unique whitebeams, found in granite gorges on Arran. However, Scotland is a stronghold for mountain plants found elsewhere in the world at high altitudes, and has alpine plants found at sea level. It is naturally important for heather, which is not found in any other country on such an extensive scale.

There are also families which do not cross-breed, but which have formed rare species by a series of mutations. As a result, more than 100 of Scotland's 160 hawkweeds are endemic, along with many 'microspecies' of dandelions and brambles. Recent studies suggest more new species are still appearing.

The main impact on plants today is human, and many woodland species have declined in abundance while others, especially grassland, heathland and arable weeds have spread.

There are signs that the rate of loss may have increased

in recent decades. More than thirty vascular plants (having tissues which conduct liquid) recorded in Scotland before 1930 have not been found since. One species thought to be extinct, however, the brown bog rush, has been rediscovered by detailed surveying.

Some of the plants which have suffered most are found in agricultural habitats. Cornfield species, for example, have fallen prey to selective herbicides and hay meadow species have succumbed to the addition of nutrients and agricultural improvement. Wetland and coastal species have been affected by drainage, fertilisation and cultivation. Even within Scotland's scarce surviving native woodlands, species such as the wood cow-wheat are still in decline, possibly because of grazing pressures. For the same reason, some high mountain plants are less prolific than they should be, with woolly willow and the alpine sow thistle on the brink of local extinctions. The decline of the blue heath, which is snow-loving, may be an indication of acid pollution or climate change.

Lower Plants

Many of these species are not proper plants, but include non-flowering mosses and liverworts and algae, slime moulds and lichens. Our knowledge of some – mosses, lichens and some algae – is better than others. Fungi are difficult to study in detail because of their seasonal nature. These species are particularly important in Scotland, which has over 80 per cent of Britian's lichen and bryophyte (mosses and liverworts) flora, representing 40 and 65 per cent respectively of the European flora. These species are often very important to the functioning of other organisms and habitats, such as peat bogs, Caledonian pine and birch woods and lichen-rich machair grasslands. They are also very important indicators of air pollution.

4
.

Fresh Water

.

No one knows the value of water till he is deprived of it ... I have
drunk water swarming with insects, thick with mud and putrid
with rhinoceros urine and buffaloes' dung, and no stinted
draughts of it either.

David Livingstone, explorer, *Private Journals*

Scum Sunday
.

MAGNUS MAGNUSSON, the chairman of Scottish Natural
Heritage (SNH), called it a bad day for Scotland. The
day was Sunday, 14 June 1992, the day on which Loch Leven
in Fife turned blue-green. Coincidentally, it was also the day
that SNH chose to have an environment conference on the
shores of the loch.

Loch Leven is a valuable wildlife site, protected by
national and international conservation designations in
recognition of its wildfowl, wader, plant and insect populations. It is home to the pink-footed goose, the mallard,
tufted duck, gadwall, shoveller and teal. It was embarrassing,
therefore, for Mr Magnusson and friends to find the water
of the ancient trout fishery covered in a toxic scum caused
by decades of man-made pollution. Notices had to be put up
to ban bathing, and visitors were warned to keep their
children and pets away from the algae.

Fishermen, meanwhile, were having a poor time. Visibility
in the water was down to half a metre, and the brown trout
could not see the artificial flies they were being offered.
Consequently, they were failing to bite. All this on a loch said
to bring around £1.5m a year to the local economy through

tourism and recreation – in other words, through the specialness of the environment.

The same thing had happened before, and would happen again in the loch's history, but environmental improvements are often inspired by single incidents, and this time the incident was serious enough for action to be taken. The algal bloom had been caused by the huge quantities of phosphorus lying in the loch sediment. The chemical encourages plant growth on land and in the water, and was dumped in the loch for two centuries in the form of discharges from a woollen mill, sewage from housing, and phosphorus-rich soil particles washed from the fields of eighty farms. The accumulation was aided by the fact that Loch Leven is in the bottom of an attractive bowl, with artificially fertilised croplands tilting into it from every direction. The town of Kinross, with its woollen mill, sits on its edge.

These sediments were released in 1992 by the extended period of calm, hot weather. The phosphorus was then taken up by algae which used the sunlight to reproduce at a fantastic rate, until the point was reached, according to Magnusson, at which the water turned as green as pea soup. The day was nicknamed 'scum Sunday'.

The algal bloom floated on the surface, blocking out sunlight and inhibiting the growth of the aquatic plants on which insect larvae, and, in turn, fish and some birds depend. It also polluted downstream waters. One of the immediate effects was the demise of the famous trout fishery; Loch Leven brown trout have been introduced to lochs and lakes all over the world because of their famed fighting qualities. The incident also served to illustrate the paucity of our knowledge about the state of our freshwater fish stocks. No-one knew how healthy the trout stock was in one of Europe's best-known game fisheries.

The owner, Sir David Montgomery, whose home stands on the northern shore of the loch, responded by introducing tens of thousands of North American rainbow trout, the quick-growing quarry of choice in many lochs and reservoirs in central Scotland. The rainbow – which is more readily caught in such conditions – is a non-native, and we are not yet certain about its effects on native stocks, but the fishing had to go on because of its commercial importance. To many purist anglers, who deprecate the widespread intro- duction of the rainbow, the move was unforgivable.

From the fishery's point of view, the first step was to deter- mine whether the brown trout had been affected. Scottish scientists, it emerged, were not used to such calculations, so a team was brought from the Republic of Ireland, which had been dealing with algal blooms for some years. I went out with the team on their first day to lift the nets they had set, and watched them pull out thousands of healthy brown trout. They found, however, that one or two year classes – fish of a certain age – were missing, which suggested the trout stock had been damaged at some point.

To deal with the wider issues, a partnership was set up in- volving SNH, the Scottish Environmental Protection Agency (SEPA), the local authorities, the landowner, farmers and the Scottish Agricultural College. Its remit was to draw up a catchment management scheme, which should lead to a healthier environment. It was decided at an early stage that it would be impossible to suck or dredge the tonnes of phos- phorus from the bed of the shallow loch, and, therefore, the solution had to be long-term and involve drastic reductions in the input of phosphorus and nitrogen.

Attempts had been made to reduce the phosphorus load since the 1970s. The single biggest improvement was made in 1987, when the Todd and Duncan mill stopped using phosphorus compounds in its manufacturing process. Since the establishment of the partnership, around £5m has been spent on direct action and research. Work was already underway at the main sewage plants, three of which have since been rebuilt or had phosphorus scrubbers fitted. Since 1987, the pollution from point sources has been reduced by 50 per cent.

The more complicated task for the partnership was the capture of diffuse pollution, mainly from agricultural run- off and soil erosion, but also from rural septic tanks. Erosion is exacerbated by the fact that modern cropping patterns typically leave much of the ground bare during the winter.

Historically, farmers were advised to put phosphorus on their land to assist growth and to improve the soil's fertility. On the best land in the catchment, specialist arable units produce high-value potato and vegetable crops, as well as cereals and oilseed rape. To the north-east, it was found that the more intensive the nature of the cropping, the more soil erosion took place, and the more phosphorus was added to the burns draining the area.

New studies will assess the level of phosphorus in the soil on each farm unit, to determine how much fertiliser needs to be added each year, if any. In some places, changes in cultivation systems, with the land being left in stubble over winter and ploughed in the spring, could ease the erosion. Run-off can also be prevented by adding straw to the soil and by building 'diversion terraces'. Some farmers applied for grants under the Countryside Premium Scheme, introduced in 1997, to help them create buffer zones between field and fresh water.

There are also issues of development and planning. The impact of any potential housing development within the catchment area has to be carefully considered, while the use of sewage sludge or animal waste as fertiliser on the land – which could increase after sewage sludge dumping at sea is banned in 1998 – will have to be controlled.

Loch Leven teaches us that environmental issues can be highly complex. For around two centuries, phosphorus was being added to the loch in significant amounts. It could take as long for the remaining load on the loch floor to be flushed out. Its story is a useful reminder of the long neglect which our rivers and lochs – so much a part of what makes this country special – have suffered.

The Resource
·

Some 2 per cent of Scotland is covered by water, which is replenished by an annual rainfall of around 1430mm. We boast more than 90 per cent of the volume, and 70 per cent of the surface area, of fresh water in the UK, but with just one fifth of the population density of England to affect the purity of the resource. There are around 31 500 lochs and 6600 river systems, with the Western Isles boasting the highest number of standing waters. The majority of Scottish lochs were sculpted by glaciers and are relatively small, with only 189 greater than 100 hectares in area. The west and north-east Highlands have the highest number of running waters, most of them short and steep, and the main rivers, the Tay, Dee, Don, Forth and Tweed flow to the east coast, reflecting the overall topography. The Tay carries more water than any other river in Britain, with an average flow of 164 cubic metres per second. In the devastating January flood of 1993 it was carrying, or rather spilling, 2269 cubic metres per second.

Our most famous loch, Loch Ness, contains almost one fifth of the total volume of standing water in Britain, and nearly twice the total amount of standing water found in all of England and Wales. Windermere, the largest lake south of the Border, would rank only fourteenth in Scotland. Then again, if the entire Scottish resource was put into one single basin, it would rank ninety-third in the league of the world's largest lakes, and the Danube alone carries more than twice as much water as the combined flow of Scotland's rivers. We are world-beaters, however, in terms of depth. Only ten lakes are known to have a greater mean depth than Loch Ness.

Our freshwater ecosystems are also of international importance for plants and animals. Most Scottish lochs are distinguished by the fact that they are nutrient-poor, acid waters which lack disturbance and are not seriously polluted. The acidity is imparted mainly by acid rocks in the uplands, often overlain with acidic glacial deposits. The naturally acidic nature of Scottish waters means they are sensitive to changes in chemistry, particularly to additional acidification from airborne pollution caused by distant industrial activity and traffic.

Lochs which are not nutrient-poor tend to be found in the lowland, eastern areas where softer, sedimentary rocks dominate and dissolve to add minerals to the rivers. Loch Leven and the Lake of Menteith are examples, but nutrient-rich lochs are also found on the machair land of South Uist.

Important waters, from a conservation point of view, are recognised by SSSI designation, but most of the fresh water resource is not formally protected. Only 17 per cent of Scotland's 1300 SSSIs contain running waters. The figure suggests there are large gaps in the protection of our river systems. More than twenty areas have been designated as Ramsar sites – wetlands of international importance – but most of them have been listed because of their birdlife. Three areas have been put forward as Special Areas of Conservation under European law.

The historic use of lochs and rivers has left behind crannogs and ritual deposits, while the main uses today are fisheries, fish farming, drinking water, agricultural and industrial water supply, the dilution of industrial and effluent discharge, hydro power, recreation and nature conservation.

The addition of phosphorus is the most significant man-made alteration to the quality of Scottish fresh waters. Under our four-step chemical classification, only 3.2 per cent of Scottish non-tidal rivers and canals were polluted in 1989. The investment in effluent treatment has produced major improvements since the 1970s, but poor-quality, class 3, and grossly polluted, class 4, waters are still found in the Clyde and Forth areas. On the down side, water quality is based on parameters which do not take account of acidified upland waters, so waters in Galloway and the Trossachs which fail an EU directive standard, are regarded as ex-cellent in the Scottish statistics. Most Scottish rivers and lochs are naturally acidic, with a pH of between 5.5 and 7.0, but up to 10 per cent of fresh waters have been reduced to a pH of less than 5.5 by acid deposition over the past 100–150 years. The deficiences in the current chemical classification also mean that lochs suffering from nutrient enrichment – from agriculture or sewage – can be given a good reading.

Groundwater

Scotland has so much surface water that only 3 per cent

of its public water supplies come from groundwater. The resource is most important for private supplies and, increasingly, for irrigation and for businesses trying to avoid charges from the water authorities. If the trend continues, it could lead to over-abstraction. Relatively little is known about the country's groundwater aquifers, which can be polluted by acid water from abandoned mines, contaminated land sites, leaking sewers and industrial sites and, locally, by septic tanks and sheep-dip disposal. Nitrate, principally from fertilisers, is the main pollutant.

Running Waters

The total length of rivers running through Scotland has been estimated at nearly 51 000km. Rivers define Scotland's identity in terms of salmon angling and whisky distilling and, by and large, they are fast and clean. They tend to show a south-west–north-east alignment, following the geological faults, and were probably established 30 million years ago. They have been altered since by periods of glaciation, which may have changed their courses and shaped the land around them.

Scotland is blessed with a surface water flow of 16 000 cubic metres per person per year, compared to the European average of 4600 cubic metres. The amounts may be difficult to visualise, but the difference is impressive! The average, annual temperature of river water throughout the country ranges from 8.5 to 10.38 C.

Since the mid-1980s, rivers have been classified by chemical sampling, measuring organic pollution, and biological sampling, measuring toxins and occasional sources of pollution. It would be virtually impossible, however, to put some class 2 rivers, the second best class, back into pristine condition because of their arable and urban catchments. Pesticides have been detected in 10 per cent of drinking water sources in the UK at very low levels.

Standing Waters

Freshwater lochs and reservoirs have a plethora of uses, including power generation, drinking water supply, angling and fish farming. According to a 1991 report, the capital value of lochs for fish farming, angling and water supply exceeded £1 billion a year. Nevertheless, there are no national reviews of loch water quality. One survey of 173 of the largest lochs in 1995 found that 82 per cent were unaffected. The remaining 18 per cent – 31 lochs – were significantly altered, with three being classed as seriously degraded. The majority had been enriched by nutrients. Eutrophication – the process of nutrient enrichment by nitrates and phosphates which can cause algal blooms and deplete oxygen levels – was identified in 1947, and became a major issue all over the world in the 1960s and 1970s, partly because of the introduction of synthetic detergents. The four major sources of phosphorus leading to eutrophication are forestry fertilisers, fish farming, sewage effluent and organic agricultural wastes.

In addition to its natural lochs, Scotland has more than 130 dams over 15m in height, ranging in size from a few hectares to the 32 square kilometes of Loch Shin in Sutherland.

Freshwater Marshes

These are not often considered by the public, but are an important part of the aquatic environment and can be home to large numbers of rare species, invertebrate fauna and breeding and wintering birds. Good examples are the Loch Lubnaig and River Endrick marshes. Standing reservoirs are also of natural heritage interest. Loch Doon, Gartmorn Dam and Gladhouse Reservoir all support valuable communities of fish, plants and birds. The value of a reservoir is generally determined by the size of the daily draw-down of water.

Our Fresh Waters Today

With our fresh waters, as with our air, things are better than they have been. Before 1800, most rivers and lochs in Scotland were relatively pristine, but after five decades of pollution the rivers had become little more than open sewers. Action was taken by Parliament following the hot summers of 1858 and 1859, when the odours from the Thames – a sewer for three million people – infiltrated the very corridors of Government. The result was a Royal Commission which studied every source of pollution north and south of the Border. The commissioners found that the subject of river pollution 'could nowhere be better studied than in the neighbourhood of Edinburgh because of the curiously specific character of the foulness which the streams and running waters of Mid Lothian severally experience'.

Some of the key polluters were paper mills, and the River Almond was also affected by the distillation of oil-bearing shales to produce paraffin. The Commission was told the river at Cramond was useless for all purposes. The Clyde, of course, was worse than the rest, with the fresh waters above Lanark flowing into a 'foul and stinking flood'. The river in Glasgow city centre was said to be 'loaded with sewage mud, foul with sewage gas and poisoned by sewage waste of every kind – from dye works, chemical works, bleach works, paraffin oil works, tanyards, distilleries, privies and water-closets'. Hmm!

The pollution was blamed on urbanisation and the growth of manufacturing industries. The population of Lanark-shire, for example, had trebled in fifty years. One of the few clean rivers at the time was the Tay, which, even in Perth, was largely unpolluted because of its great volume of water.

The Commission led to legislation, but Scottish rivers continued to deteriorate for another fifty years. In 1927, 880 polluting discharges were recorded in 223 river basins, of which only 78 received satisfactory treatment. And it was not until after the Second World War that the government saw that new legislation was long overdue. The Rivers (Preven-tion of Pollution) (Scotland) Act was a turning point in water pollution control. It set up the independent, catch-ment-based river purification boards and made it a criminal offence to discharge sewage or trade effluent without con-sent. River inspectors were appointed, monitoring schemes set up and the first regular biological surveys of the Clyde began, believe it or not, only in 1968, when no fish could survive within the Glasgow boundary. By the 1980s, the Atlantic salmon had returned.

A further Act in 1965 tightened the earlier legislation and allowed the boards to take action over long-established discharges. Since then, steady progress has been made. Today it is possible that 10 per cent of our rivers and lochs are degraded to the extent that the original flora and fauna are largely or completely missing. This level of alteration is made up of a large upland sector affected by acidification, and a smaller lowland component affected by poor water quality. The remaining 90 per cent of waters are in very good condition.

Official statistics do not take account of all forms of pollution and show that 97 per cent of all Scottish fresh waters are in good condition, with only 0.1 per cent being 'grossly polluted'. However, these figures are based on the length of river affected, not the volume of flow. Therefore, a polluted 10km stretch of the Clyde is regarded no differently in the statistics from 10km of the smallest Highland river. Clearly, this skews the results to exaggerate the healthy state of Scottish waters.

All things considered, Scotland's fresh waters are a relatively healthy resource and, at least in central Scotland, are in much better condition than they were for the first half of the twentieth century. But the public continue to express concern about water pollution, which, like transport, is an issue with a public face. A walk by the unfragrant River Almond at almost any point along its course would illustrate the problem. Before the millennium, the effluent from the numerous sewage works along its banks should be diverted to a treatment plant in Edinburgh. For now, it remains one of the dirtiest rivers in Scotland, with a flow in the summer months that is up to 50 per cent sewage effluent.

The history of the River Almond is a useful case study. It is a river which could not cope with the demands placed on it by the new town of Livingston and residential and business development west of Edinburgh. It is turned orange by the iron deposits from leaking mine water near Harthill, flows black or purple, often flecked with foam from a detergent works, in Almondell Country Park, and emerges gasping for oxygen into the Forth at Cramond. Not surprisingly, it is a key priority for East of Scotland Water.

Other freshwater problems, some of them mentioned in Chapter 3, include the introduction of non-native fish; algal blooms caused by nutrient inputs and unco-ordinated and unscientific management by riparian owners, who rarely consider the health of the whole river. It is also unfortunate that the former river purification boards, replaced by SEPA, had to concentrate on river pollution at the expense of our lochs. There is currently no strategic management or monitoring of fresh water lochs.

Peter Maitland, a fisheries scientist, would like to see twenty-five catchment-based regional fishery boards and trusts being established to prevent further losses of fish stocks and damage to habitats. He has suggested they could be funded by local initiatives, contributions, angling permits and a national angling licence, which has proved unpopular in Scotland but has worked in England and elsewhere.

Man's inputs are many and varied, but three of the current and most important problems are pollution from sewage, polluted water from abandoned mines, and urban drainage. Pollution from industry is dropping due to regulation and the restructuring of Scotland's industrial base, and sewage is being subjected to higher treatment standards. But urban run-off is increasing as traffic increases, while there are just two remediation schemes in place to deal with mine drainage from 170 abandoned sites.

Other important issues include acidification caused by polluting emissions from traffic and industry, the use of fertilisers and pesticides in farming, the effects of hydro schemes, and the predominance of monocultures of commercial forestry or sheep grazing in upland areas. Plantation coniferous forestry is an important source of phosphate run-off. It also gathers acid pollution from the atmosphere, via its needles, and releases it into delicate freshwater ecosystems. Sheep farming, meanwhile, often combined with estate management for deer, has kept native forestry cover at minimum levels and has contributed to flash flooding which causes erosion and the loss of topsoil. Although fertiliser and pesticide use is declining, the system is already so over-loaded that the changes will make little difference.

Most of the issues mentioned now come under the remit of SEPA, which was set up with great good intentions, and with an enthusiastic and committed workforce, but which is suffering from serious under-funding, while the industries which it regulates are worth huge sums of money. It also has to deal with air pollution, waste management and the marine environment. Its last budget was only £24m.

In recent times, the big improvements in water quality have been driven by Europe, through the Urban Waste Water Treatment Directive which deals with sewage, and the Nitrates Directive which tackles pollution from agriculture.

The Issues

Sewage remains the single biggest cause of poor water quality, due to inadequate treatment and overflows from sewers which also carry street drainage.

The Scottish population produces around 1.1 million cubic metres of liquid domestic sewage very day, containing 110000 tonnes of solid matter – with the potential to cause widespread damage. The growth of bacteria, fungi and algae, which use the organic matter in sewage, can result in slime on a river bed and the loss of life-sustaining oxygen in the water.

The more resistant sanitary wastes – which should not be flushed away in toilets, but put in bins – end up on river banks and beaches. Some pollutants in sewage, such as heavy metals from trade effluent entering sewage plants, are toxic to fish life. The natural environment can cope with a certain amount of sewage, but the equation depends on the degree of dilution in the receiving river.

Over 95 per cent of Scotland's population is connected to the 1450 sewage treatment works, and because the population is concentrated in coastal areas, 48 per cent of sewage effluent is discharged to the sea after screening for heavy solids and settlement. Discharges to estuaries account for 28 per cent, and 24 per cent goes to fresh water, where secondary (biological) treatment renders the product cleaner. One flaw in the system is the fact that there are 2000 sewer overflows in Scotland which operate when there is heavy run-off from rainwater in the streets and can release raw sewage into rivers and on to beaches. Many operate more often than is intended due to overloading caused by new housing or structural problems.

Not surprisingly, we lag behind other countries in our treatment of sewage. In Sweden and Switzerland, more than 90 per cent of sewage treatment works include phosphorus removal, compared to just three works in Scotland.

Urban Drainage and Industry

Urban drainage is a newly recognised problem, but is becoming more acute. Surface water suffers from chronic contamination from heavy metals, oil and rubber from road drainage, the deposition of particulates and hydrocarbons, air pollution and car-washing detergents. This results in the downgrading of 20 per cent of class 3 and 4 rivers. The worst incidents are found around industrial estates and business parks, though the impact of industry has reduced in the past twenty years, partly due to the loss of heavy industries and the spread of the food, paper, petrochemical and electronics sector, which have less polluting processes.

The more water an industry uses, the more effluent it produces. The whisky industry, for example, uses 1415 tonnes of water to produce 32 tonnes of spirit and 1248 tonnes of highly polluting effluent.

Of Scotland's 85 distilleries, 32 have biological treatments on-site and 28 transfer their waste for treatment off-site. The rest put the waste on to land, or discharge it into rivers and estuaries.

The paper industry also produces large volumes of liquid waste and potentially harmful chemicals. Secondary treatment has been installed in several big plants in the past ten years, but paper mills have been fairly regular offenders.

The petrochemical industry has a wide range of persistent and toxic substances to dispose of, although investment in environmental protection is substantial. And the booming electronics business, employing 54000 people, uses potentially hazardous substances, including acids and solvents, but has tended to adopt a pro-active approach, and was established at a time of heightened environmental awareness.

In many areas of central Scotland another potential threat comes from abandoned mines which leak iron-rich pollution and affect 22 per cent of the lower class rivers. The scale of the potential problem is indicated by the fact that, in the mid-1800s, Scotland had 560 active coal mines, compared with two today. Mining is expanding again, through opencast operations. An alliance of environment groups called, in May 1997, for all opencast mining applications to be called in until a policy review could be conducted. And a Labour Party commission described opencast mining as 'one of the most environmentally destructive practices in Scotland'. Dozens of communities in the central belt have fought against new developments, although permission has been given for a number of large mines, including one at Glentaggart, South Lanarkshire, which will extract 10 million tonnes. The impacts include noise, dust, water pollution and loss of amenity. Opencast developments are also major scars on the landscape, unlike deep pits.

The discharge of pollutants, including iron, from old and new mines can turn streams orange, destroy spawning grounds, kill invertebrate populations and cause gill damage in fish. When mine water broke out of the Dalquharran Colliery in 1979, fish life was destroyed over a 15km stretch of the River Girvan, from the colliery to the sea. The situation may be improved from 1999, when it will be possible to take legal action against former mine owners or operators.

Agriculture

Agricultural output from Scotland's 18000 full-time units increased steadily over the thirty years to 1980, with some decrease since due to moves to reduce production. The industry accounts for less than 3 per cent of GDP, but is responsible for more than 50 000 jobs, which is more than any single manufacturing industry. Only 5 per cent of Scotland's most polluted stretches of water are affected by agriculture, which is far more important for the damage it does to the class 2 rivers. Intensification of agriculture has had two main impacts, with up to 700 incidents a year involving pollution from farm wastes and water abstraction. Phosphorus sheep-dip contaminants have been responsible for serious short-term incidents in some places, and fertilisers, particularly nitrogen and phosphorus, have leached from the land. In many areas of the Highlands, intensive sheep and deer grazing has promoted river bank erosion which has produced wide, shallow channels which are inhospitable to large fish.

During the summer months, many pollution incidents are caused by silage discharges, and in winter pollution is often in the form of run-off from waterlogged fields following the spreading of slurry. Effects on water courses are likely to be most serious in areas of intensively cultivated arable farming. The River Eden in Fife, for example, has elevated nitrate levels along much of its length from Glenfarg to Guardbridge, near St Andrews.

Water Abstraction

Abstraction is used for drinking water, irrigation and hydro power. Good-quality water has always been prized: in 1804 a Mr W. Gibb delivered treated water to the centre of Paisley to sell it for one halfpenny a gallon. In Glasgow, the major typhus epidemics of 1847, 1848 and 1853 led to the Glasgow Corporation Water Works Act of 1855 and the promotion of the scheme which led to the city's water being taken from Loch Katrine. Successive Acts of Parliament at the end of the nineteenth century gave powers to local authorities to arrange public water supplies, so that by 1914 most towns and cities generally had good supplies. There are 560 public water abstractions, but there is no record for private abstractions.

The key health issue after the establishment of public

supplies was the use of lead pipes, and the fact that Scotland's upland water was plumbosolvent. Lead can cause brain damage in young children, but lead piping was only abandoned in domestic plumbing as recently as the 1950s. It is still being removed with the help of subsidies.

Glasgow takes its water from Loch Lomond and Loch Katrine; Aberdeen depends on the River Dee, and Edinburgh is supplied by the Megget and Talla reservoirs. Some land-based fish farms take large volumes of river water, therefore reducing the flow and affecting migrating fish. SEPA has no control over the volume of water taken.

Hydro Power

Hydro power has greatly altered catchment hydrology in the Highlands over 100 years. The early schemes, from 1895 to 1930, were created to provide electricity to smelt aluminium, while the power schemes were commissioned between 1950 and 1962, with pumped storage schemes being developed later at Cruachan and Foyers.

Hydro power has altered the hydrology of perhaps 20 per cent of the catchment area. Lochs have been expanded by dams and new reservoirs have been created, although the total surface area of new water is small. There are 65 power stations and, with the exception of Cruachan and Sloy, they are usually storage systems in which water flowing from a catchment is regulated by a reservoir. The pumped storage schemes recycle water by pumping it back uphill after it has passed through the turbines. By the mid-1960s, the most attractive catchments had been exploited and there was a switch to the larger, pumped storage systems.

The ecological effects involve disruption to benthic (bottom-feeding) flora and fauna in lochs and intercepted streams. Salmon migration is impaired in many locations by barriers and by water mixing from catchments, and spawning redds have been lost. However, fish ladders or locks were incorporated during dam construction to allow migrating fish to reach their spawning grounds, and studies in some areas have indicated that fish catches have not altered breeding success. For example, salmon catches were not found to have changed in the two decades after the construction of the barrage on Loch Awe. It is harder to assess the effects of hydro power on downstream areas which are affected by changing water flows from the dam. In some of the most scenic areas, the variation in water levels is kept at a minimum – on Loch Faskally at Pitlochry, for example, and on Loch Benevean in Glen Affric.

No conventional hydro schemes have been promoted since 1965, but the growth of renewable energy has led to the construction of small 'run-of-river' schemes in which short, steep falls are utilised with an intake weir and small headpond, a short underground pipeline and a small power station housing one or more turbines.

Salmon, Sea Trout and other Fisheries

The management of fish populations has been piecemeal and unscientific, with a strong bias in favour of salmon, and no integrated national structure. The angling potential of Scotland was first publicised in the seventeenth century by Richard Franck, a soldier with Cromwell, who wrote about his experiences. In the late eighteenth and nineteenth centuries, angling in Scotland was described at length by sportsmen, tourists and naturalists. Trout angling developed in the nineteenth century, with numerous sporting hotels introducing new trout stocks from hatcheries.

Unfortunatlely, owners have often carried out management practices which have damaged the resource. These include poisoning the water to eliminate unwanted pike or perch, netting to remove unwanted fish, electro-fishing, the introduction of 'food species', raising the water level, adding fertilisers and blocking inflows or outflows, creating pools and spawning areas and constructing groynes and fishing jetties.

In fact, pike can play a beneficial role by eating young fish with a naturally high mortality, thus reducing the dense populations of small fish, and eating diseased fish and stocked fish.

Stocking has rarely been scientifically based, but was motivated by a desire to add new blood, or increase the number of fish caught. In some cases it may have introduced diseases or caused serious damage to the established ecosystem.

Native salmon and sea trout have been prized since the late middle ages, and were barrelled and shipped to England and continental ports in the late eighteenth century. The main method employed was the beach seine net, which is still in use today. Until the 1980s, there were net fisheries in the estuaries or lower reaches of most of the big salmon rivers, but the increase in salmon farming led to a drop in commercial netting. A trust set up by angling interests to buy off the salmon nets has led to the removal of these fisheries on the most important salmon rivers, including the Tay. Salmon fishing can cost anything from £10 a day to more than £1000 a week, and the development of time-sharing on salmon rivers – where an angler buys one week in perpetuity – has inflated capital values to as much as £15 000 for each fish.

The eggs, parr and smolts which form the developmental stages of juvenile salmon prior to migration are sensitive to a wide range of factors, in particular habitat damage, acidification and predation. On many rivers, lack of management in the headwaters – which are not valuable for fishing – has had an effect on the fishing. These include the collapse of banks caused by grazing livestock, the obstruction of streams by trees and other debris, and the abstraction of water for irrigation.

There have been long term declines in fish catches on the Beauly, Conon, Alness, Dee and Forth, and the big multi-sea winter fish – which have completed their migration and breeding cycle more than once – are rare. In recent years a catch and release policy has been encouraged on some rivers, notably the Little Gruinard in Wester Ross, for conservation reasons. The Scottish game fishery is valued at around £1000 million, with an annual revenue generation of over £50 million in 1989.

There was once a net fishery for Arctic charr on Loch Leven, but the species became extinct in the 1830s, either because of the fishery or the lowering of the level of the loch. There has never been a serious fishery for eels in Scotland, but fishermen from England and Europe have been using fyke nets to catch them in recent years.

Up until the 1970s, there were several itinerant professional pearl fishers in Scotland pursuing a trade which thrived in the fourteenth century.

Fish Farming

The farming of rainbow trout began in the 1960s and was modelled on the established Danish industry. It is now worth several million pounds a year. The effects of feeding, waste and chemicals can cause nutrient enrichment, fungal growth and the introduction of chemicals with uncertain long-term effects, though the impacts are thought to be largely confined to the immediate vicinity of the fish cages. There have also been numerous escapes – sometimes when nets and cages have been cut deliberately – which have released tens of thousands of rainbow trout in places including Loch Tay and Loch Awe. The rainbow trout in the wild are a bonus of sorts for fishermen, but may have damaging effects, in terms of competition, on brown trout. There has been a significant expansion since 1980 in Atlantic salmon smolt production, for transfer to salmon farms at sea, with more than 20 million smolts produced annually.

Public Water Supply

Public water supplies remove something like 1.2 per cent of the daily discharge of Scottish rivers, with reservoirs and lochs supplying 85 per cent of the total, and rivers supplying 12 per cent. Around 355 lochs and reservoirs – 1 per cent of the standing waters – are used for water supply.

Acidification

Scottish waters showed no change in acidification for 10 000 years, until the effect of the industrial revolution took hold in the mid-nineteenth century. Most Scottish rivers and lochs are naturally acidic, which makes them vulnerable to pollution carried by rain, snow and moist air. Around twenty to thirty lochs are believed to have become fishless and many more streams have been affected by deposition of acid pollution in rain, mist and snow. Coniferous forestry is also responsible for 'occult' deposition in which the tree needles act as a sponge and gather acid droplets from the atmosphere. High levels of toxic forms of soluble aluminium occur in waters with a permanent pH below 5.5. As acidity increases, changes take place in fish and invertebrate populations, with reductions also noted in some bird populations. High mortalities of fish eggs and fry have been reported in polluted waters with a pH less than 5.5, and an absence of adult fish generally indicates very advanced acidifcation.

Acid rain has been a fact of life in Britain since the start of the industrial revolution, but it was not until the 1960s that it was learned that sulphur dioxide (SO_2) could be carried great distances, and that Scandinavian countries were being damaged by substances from the UK and central Europe. Lochs and rivers vary greatly in their sensitivity to acid deposition, depending on their buffering abilities and the amount of calcium in the soil. Their damage threshold is known as the 'critical load' which they can accept. These critical loads have been worked out for lochs throughout Scotland to establish what action needs to be taken to prevent serious acidification.

The worst-affected areas are Galloway, the Trossachs, North Arran, Rannoch Moor and the Cairngorms. Acid deposition is lower north of the Caledonian fault, due to the distance from pollution sources and the direction of the prevailing wind. In the early 1990s, slight improvements in water quality were recorded at Loch Ard near Aberfoyle indicating a decline in the sulphate content of rain.

Not much is known about acidification impacts on upland rivers, but it is probably partly responsible for long-term salmon catch declines in several fishery districts.

Sulphur outputs have been reduced from coal-fired power stations by 80 per cent, but scientists at the Macaulay Land Use Research Institute have estimated that lochs in Galloway and the Trossachs will still contain sensitive waters even after SO_2 emissions have been reduced by international agreement. Nitrogen levels are thought to be increasing because of the growth in traffic.

The Department of the Environment (DoE) has published studies on critical loads and projected declines in the emissions of sulphur up to the year 2005, which suggest that nine 440 square kilometre areas south of the Caledonian Fault will still be receiving damaging acid rain five years into the next millennium.

Afforestation

Afforestation, particularly where dense commercial plantations are introduced, can have serious effects on the hydrology, chemistry and biology of fresh waters. These were not fully understood until well into the 1980s. Fresh waters have been affected by the considerable growth in afforestation. The Forestry Commission argues that these impacts have not been extensive, and have generally been related to instances of bad practice, which have mostly been eradicated.

Many of the problems have been caused by downhill ploughing – the ground has to be broken up to allow the trees to take root effectively – which allows soil and fertilisers to be carried into water courses. It is often too dangerous to plough along the contours, and therefore intercepting ditches are created to control the run-off. Erosion is greatest immediately after ploughing, with the next major impact thirty years later when the trees are harvested and the ground laid bare again. Most planting in Scotland has been in upland areas which have little agricultural value, and so nutrients such as phosphates, nitrates and potassium are applied at the time of planting and when the trees are six to eight years old. The usual fertiliser is rock phosphate, around 10 per cent of which can leach into drainage water in the first year after application. There have been a number of instances of algal growths after forest fertilisation. The Forestry Commission now issues advice on measures which can be taken to ameliorate the effects, such as fertiliser only being applied after a soil nutrient assessment, and being spread by hand in sensitive areas, with buffer zones being established along watercourses.

Trees, of course, are thirsty, and up to 1.5 per cent of a catchment area's annual rainfall can be lost for every 10 per cent of forest cover. In dry conditions, the overall catchment yield may be reduced by 20 per cent. The ploughing and draining of wet soils can also lead to increased summer flows, and the erosion of ploughed hills can generate sediment which can damage spawning redds.

The commercial conifer of choice in Scotland is the American sitka spruce, which is favoured because it has adapted to thrive in nutrient-poor soils by obtaining nutrients from rainfall through 'ion exchange' on the needle surface.

Recreational Impacts

The increasing importance of recreation in modern society has led to pressure on fresh water lochs through the development of leisure facilities and holiday homes. In some popular areas the wear and tear caused by visitors is obvious, with severe damage in evidence at Loch Lomond, where disturbance to wildlife, bank erosion and litter are common.

5

·

Marine Environment

·

Du's beauty an freedom and life an death,
An we sood aa staand a-daar a dee,
For nae man can tame dee an nae hadds keep,
Bit du's inspiration ta me.

Rhoda Butler, Shetland, *Da Sea*

The People of the Sea

·

THE NATURALIST FRANK FRASER DARLING once waxed
anthropomorphic and remarked that seals were the
'people of the sea'. He should have known better, but his
view is now common currency. Shetland poets over the ages
have more accurately observed that Shetlanders are 'people
of the sea', and often live and die by it. Fishermen, however,
appear to be less important than seals in the reckoning of
the British public. The last seal cull, carried out in Orkney in
1978, was a public relations success for Greenpeace, and a
disaster for islanders. Television pictures showed baby seals
being bludgeoned to death as their watery-eyed mothers
looked on. In fact, seals were not clubbed in Orkney and
Shetland – the technique was common in Arctic Canada –
but were shot. They were shot because fishermen believed
the population had to be controlled. They argued that
the seal population was out of balance with the rest of the
marine environment, and was consuming vast quantities of
fish at the expense of the real people of the sea.

Whether they were right or not, the culls were stopped
for the wrong reasons. Seals are mammals, seals look good,

people in cities like the look of seals, ergo: they should not be killed by fishermen intent on raping the sea.

A senior marine scientist once told me: 'When you are in the middle of London you can afford to have a romantic view of something that is distant from you. People in London would give up fish rather than have seals killed. They would sacrifice the Scottish fishermen before the seal pup.' Twenty years on, decisions are still being determined by public opinion. Fishermen in Scotland and Ireland wanted to talk about the seal population in 1994, and made reasonable approaches to Scottish National Heritage (SNH) and the Scottish Office. But the media learned of the impending meeting, and the environment agency cancelled it, not because it was in possession of scientific evidence to prove that a cull was unjustifiable, but because it did not want bad publicity. The incident says a lot about the muddled state of environmental understanding in the UK.

There is a widespread acceptance that there are too many red deer in Scotland, and even animal rights organisations accept the need to kill tens of thousands of animals every year. Yet deer, like seals, are mammals, and they have nice eyes too. Certainly, we understand the terrestrial environment far better than we understand the marine environment. There are fewer variables, and we have messed up the land to a far greater extent than the sea. We can also see what is happening. We know what deer eat, and we know that too many deer equals no trees and serious erosion. We are not so certain about the seal's diet, or its effect on its environment. Here is another failure of past governments: scientists have been refused funding for seal research projects for purely political reasons. If the proposed study was likely to conclude there were too many seals, the application was turned down.

Meanwhile, fishermen feel they are being increasingly squeezed by legislation and competition. Competition from the seals must seem particularly unjust to some lobster fishermen in the northern isles. These men are at sea all day. They see the seals, and they find their lobster creels raided by seals, week after week. It is perhaps not so surprising that one or two of them cracked and took the law into their own hands. On more than one occasion, a fisherman has taken a shotgun to the beach to shoot a few dozen seals out of frustration and anger, and the belief that no matter what evidence was presented, the view of the middle-class yuppie was bound to prevail. Shetland is 1000 miles away from London, but light years away in terms of environmental understanding.

Such actions, of course, are unjustifiable, and may be entirely misguided. Shooting a few predators when there are 30 000 grey seals around Orkney will do nothing to ease life for the lobster fisherman. When one predator is removed it is replaced by another one. But the senseless shootings are the result of our failure to deal with the issue. Such acts also illustrate the depth of our ignorance.

Fishermen say that a seal will eat around two tonnes of fish a year, which means that the Scottish population of grey seals consumes over 200 000 tonnes annually. But birds eat far more fish, whales and dolphins and porpoises eat huge quantities of fish, and fish eat huge quantities of fish.

Research has been going on for some time in Ireland on the diet of grey and common seals, and the Sea Mammal Research Unit at Cambridge is also working on the issue, although it was its research from the 1980s which was often used to reject calls for a cull.

It found that seals ate mainly sand eels, and not the larger commercial species, but the results were based on the examination of seal faeces for small earbones. Most fishermen, and many scientists, point out that seals are inefficient feeders and will often eat only the soft parts of salmon, discarding the head – and the earbones. The lesson of the seal debate is simply this: we cannot afford to be sentimental about animals when we are part of the equation and when we have altered their habitats, whether the animals are grey seals or elephants.

The same scientist who lamented the romantic view of the Londoner suggested that a massive cull, of up to 50 000 animals, would be required to make any difference to the fish stocks available to fishermen. Slaughter on such a scale is unthinkable. On the other hand, the proper research may well prove that seal numbers are an insignificant factor in sea fish catches. We should not be afraid of knowing the answer and facing difficult decisions.

But if culling is unacceptable, there is another way. Some mammal populations have been controlled by sterilisation, and Canadian scientists are developing darts which sterilise seals in their waters. Surely that would be acceptable.

We do not, after all, become consumed with feeling for the water vole, which is fast declining because of habitat damage caused by us. And we do not wince over the fact that one of our rarest birds, the corncrake, and its chicks, are being cut up by farm machinery in hayfields every summer.

And Shetland, bear in mind, is only 200 miles from the coast of Norway, where seals are culled and used for their fine meat and skins. Even if a small cull was allowed, it would be hard to imagine seal meat on a Scottish menu.

The Resource

.

Scotland's serrated coastline and its marine environment are outstanding in national and international terms. The coast is 11 800km long, with 800 islands, more than 600 of them on the Atlantic seaboard, and an unknown number of small islands and skerries.

The total area of marine waters – taken to the 200-mile limit – exeeds the land and freshwater area eight times over and covers two major provinces – warm-temperate Lusitanian, and cold-temperate Boreal waters, with Shetland experiencing boreal-arctic conditions. Our marine environment supports around 13 000 species, ranging from mussels to minke whales, but none is endemic.

Around 65 per cent of the sea area is deep sea and 35 per cent is continental shelf, with a highly productive zone between the two regions – the 'shelf break' – which has large populations of fish and seabirds and is a migration route for whales.

The east coast is predominantly low-lying, with the main sea cliffs in Caithness, Berwickshire and along the Buchan coast. Sedimentary shores, wide firths and gentle, curving coastlines are typical of the east, which is broken up by five major firths. The north and west is complex, fjordic, highly indented and broken by hundreds of sea lochs. Coastal types vary from vertical rock faces to sheltered muddy shores on the innermost reaches of sea lochs. The three major archipelagos, Shetland, Orkney and the Western Isles, are all distinctive, each with is own coastal habitats.

While Shetland has a complex geology giving rise to a variety of cliffs and rock platforms, Orkney is composed of old red sandstone, giving a landscape of rolling hills and gentle basins. The Western Isles are mainly composed of ancient Lewisian gneiss.

The country's tidal waters are also of great economic importance and support inshore fisheries, fish farming, recreation and tourism. Most of our major industries, including power generation and petrochemicals, use coastal waters for cooling, transport and disposal of effluent. Some tidal waters have good natural cleansing abilities, but the enclosed sea lochs flush slowly and are sensitive to pollution.

As every schoolchild learns, conditions on the west coast are influenced by the Gulf Stream. By contrast, the east coast waters of the North Sea have a much wider annual temperature range, with colder winter and warmer summer temperatures. The mixing of the warm Gulf Stream and the cold Arctic waters around Scotland means that it is notable for the number of species at the limit of their southern or northern ranges. The waters of the west coast are relatively unpolluted, with the obvious exception of the Clyde, which drains the industrial heartland of Glasgow and Lanarkshire. It is one of the two major estuaries – the other is the Solway – which characterise the south-west of Scotland and differentiate that area from the north-west coastline. In the north and the north-west, the relative purity of the waters contributes to a high diversity of marine flora and fauna along the shores and in the shallows.

The east coast has much more industrial, agricultural and urban development which contributes to nutrient loading and high pollutant levels in the North Sea, which also suffers because it is encircled by industrial nations. The discovery of oil and gas in the 1960s necessitated, for the first time, detailed planning of the coastal environment, although most of Scotland's coast remains undeveloped, with rocky shores making up 50 per cent of the whole.

Cliffs and Rocky Shores

Sea cliffs, which are important for nesting sea birds such as fulmar, guillemot and razorbill, are under local pressure from agriculture, recreation and coastal developments. Some popular climbing cliffs, including the Old Man of Hoy, are littered with the ironmongery left behind by climbers.

Although kittiwakes, puffins and arctic terns dropped in numbers in the Shetland isles in the late 1980s and re-covered in the early 1990s, there are no overall trends to suggest birds associated with cliff areas are in decline.

The rocky intertidal zone can be very steep and often is home only to barnacles, limpets and northern shore wrack, while sheltered rock and boulder shores, especially those at the head of sea lochs, may be used by seals as haul-out and breeding sites and by otters which feed and make holts along the shore.

Soft Coasts

Dune and shingle beaches are important for seabirds in-cluding terns, black-headed gulls and eider duck. In many areas, dunes have been stabilised for afforestation, housing, golf courses and other recreational purposes. The beaches themselves, however, are mainly subject to light use.

Machair, a habitat formed from rich shell sand carried inshore by storms, occurs on the west coast of Scotland and the Hebrides and has been used for centuries for grazing by crofters. It can also be cultivated for oats and rye, although very few crofters till their land today. The vegetation of the machair is flower rich, with large numbers of orchids. It is also an important habitat for nesting birds, including dunlin, redshank, oystercatcher and ringed plover.

Machair has to be maintained by grazing; where grazing has ceased the coarser grasses take over. On its nature reserve in Rum, SNH reintroduced cattle to the island in order to create the right conditions for flower meadows to survive. In some areas, increased numbers of livestock have been kept on machair in winter and chemical fertilisers and herbicides have been used. The pressure of people, sheep and rabbits is causing serious local erosion in some sytems, notably on Lewis.

Scotland has limited areas of saltmarsh, which can also deteriorate when the traditional grazing activity is stopped. At Caerlaverock National Nature Reserve on the Solway, the trend towards species-poor swards has also been reversed by the reintroduction of grazing. Other saltmarshes have been used for tipping, road building, housing, and the construction of sea defences. The major areas are found in Scotland's estuaries and offer abundant food sources for over-wintering wildfowl and waders. They also act as stop-over points for migrating birds, and as nurseries for young fish.

Human pressure on these areas of coast is considerable, with increasing recreational use. Beach-cleaning machinery, for example, can remove large numbers of 'infauna', the animals which live in the sand, and disrupt the sediment structure. Bait digging, cockle harvesting and sand extraction have also had local effects. A new law was introduced in 1995 to stop mechanical harvesting of cockles on all exposed beaches.

Sea Lochs

The 110 sea lochs listed on the west of Scotland can be divided into three types – fjordic, fjardic and open. Fjordic sea lochs, like those found in Norway, include Lochs Sunart, Seaforth and Hourn, and are typically elongated, steep-sided, deep, and divided into basins separated by sills. Fjardic sea lochs, although the name is not in common usage, predominate in the Western Isles; they are broader and shallower than fjords, and often have numerous islands. Their basins are separated by narrow channels which may create strong tidal currents and support specialised wildlife, such as seapen and featherstars. Open sea lochs, like Loch Gairloch, have no sills or basins and are more exposed to waves.

Some sea lochs which are important for wildlife are also under pressure from largely unregulated commercial dredging for clams, which causes serious damage to the seabed and may be causing local extinction of species. Scotland currently has no marine nature reserves, and coastal areas have traditionally lacked attention and statutory management. In 1989, for example, there were nineteen separate developments in estuaries, half of which were for rubbish tips, and a further twenty proposed for uses including car parks, housing, barrages and marinas.

Shallow Seas

These are important for fish, seals and cetaceans, including the harbour porpoise. The only resident North Sea population of bottlenose dolphin, and one of the few in Britain, is found in the Moray Firth, which receives sewage pollution from Inverness and the surrounding area and numerous industrial effluents. Seabirds, including the gannet, patrol the shallow seas, feeding on the abundant young fish and breed on the mainland, archipelagos and isolated islands like Ailsa Craig, the Bass Rock and Fair Isle.

The sublittoral area (the seabed immediately below low water mark) is important to the maintenance of high biodiversity and is at risk from pollution. Recreational sailing, jet skiing, scuba diving and other water sports need to be managed in sensitive areas.

Kelp and maerl (a mixture of sand and seaweed) harvesting – formerly a major industry in the eighteenth and nineteenth centuries – is thriving in some areas again and damaging wildlife, including the organisms which grow on the plants or which use the beds of kelp as their habitat. In the Western Isles, the removal of kelp could also promote erosion of the machair through the loss of its damping action on the waves. The sublittoral benthos – the wildlife at the bottom of the sea – is not fully recorded.

Estuaries

The estuarine areas, the Clyde, Solway, etc., are the most populous parts of the country, with industry, military facilities, tourism, wildfowling, commercial fishing and agriculture competing for space. Scotland's forty-nine estuaries are also some of the most important areas for wildlife.

An estuary is a partially enclosed area of water where sedimentation causes the development of 'soft' tidal shores which are exposed to salt and fresh water.

Estuarine mudflats are highly productive ecosystems and may contain nutrients from decomposing saltmarsh, natural, agricultural and urban sources. On the west coast, the Western Isles and the Northern Isles, the input of nutrients is lower because of the absence of high populations and intensive agriculture. Most additional nutrients come from sewage and from agricultural fertilisers. The Clyde and Forth are regarded as highly-enriched, while the Ythan Estuary in Aberdeenshire, which drains an area of intensive farming, is polluted by nitrates and has recently been declared a nitrate-sensitive area under a European directive.

More than 95 per cent of estuarine water is classified as 'good' to 'fair', and conditions in the Clyde and Forth have improved markedly in recent decades. Historically, both big estuaries were heavily polluted, the Forth by paper mills, the petrochemical industry, intensive farming and large quantities of untreated sewage. Recent monitoring of mercury in flounders and mussels, a standard test, showed a decrease in the metal over the past ten years.

The variable concentration of salt in the water in estuaries, due to the inputs of fresh water, creates difficult environments for wildlife. This means that species diversity is low, but where wildlife has adapted to the habitat – the eider duck, for example – animal populations can be very high. In the mudflats, the invertebrate communities of worms can reach densities of one million per square metre, offering an almost limitless source of food for wading birds. Crustaceans, molluscs and bivalves, including the cockle, also form important food supplies for birds. Characteristic plants include the horned wrack, glassworts and eel grasses.

Mudflats are nursery areas for many species of fish, feeding areas for many species of resident waterfowl, and are on the flyway routes for migrating birds. The pink-footed geese and the scaup are important visitors to the Solway, barnacle geese congregate on Islay, greylag geese in the Dornoch Firth and the red-breasted merganser in the Moray Firth. Several estuaries have been designated as Special Protection Areas (SPAs) under the European Birds Directive.

Reclamation of land for development is a major threat to estuaries, and is most notable on the intertidal area of the Forth estuary, upstream from the Forth Road Bridge, where 50 per cent of the habitat has been reclaimed. Almost half the intertidal area of the inner Forth has been reclaimed in the last 200 years, much of it for petrochemical development.

Open Sea

Scotland's coastal waters reach depths of around 30m, before giving way to the open sea areas of the continental shelf, with depths of 200m and more. The open sea on the east coast is relatively shallow water when compared with the Rockall Trough off the west coast, which has depths over 1000m close to the coasts of the north mainland.

The water masses of the open sea are moved by winds and tides, and 'upwellings' replenish nutrients in the upper layers of the sea and make them available to phytoplankton and zooplankton, the microscopic aquatic plants and animals which form the base of the marine food chain. The main currents on the surface are driven by wind and the topography of the seabed.

A species called the hatchet fish is unusual for its occupation of the mid-water over the continental shelf, and its migration, 200–300m up and down, depending on the light levels. The same areas are important for the larger cetaceans, the whales, which can also be sighted in inshore waters. Little is known, however, of the distribution of deep sea invertebrates, although our knowledge of the waters of the continental shelf is increasing through the development of the offshore oil industry.

Most oil pollution incidents in the North Sea, and around the world, are caused by day-to-day sea-going traffic – some incidents are deliberate, some accidental – and by oil exploration rigs and production platforms. The wreck of the *Braer* oil tanker on the south Shetland mainland in January 1993 attracted great publicity, but the overall effect of such incidents is not as damaging as the constant, insidious leaking of oil. Bulk carriers, whatever their cargo, can also damage marine ecosystems by discharging ballast water which may introduce pollutants, or exotic species, picked up in other parts of the world.

The Issues
.

Fisheries

Fishing in the EU is regulated by the Common Fisheries Policy, which sets a Total Allowable Catch (TAC) which is divided up into quotas for each member state. Vessels must have access to a quota, and have to stop fishing when the quota is exhausted. The main stocks for Scottish fishermen are cod, haddock, whiting, shellfish and herring. The EU also sets minimum landing sizes for commercially-exploited species.

UK vessels have exclusive fishing rights out to six miles, but foreign boats have access between six and twelve nautical miles offshore. This allows boats from Germany and the Netherlands to fish off southern Shetland and Fair Isle, boats from Ireland to operate in south-west Scotland, and French and German skippers to fish around the Outer Hebrides, St Kilda and the Flannan Isles. There has been controversy in recent years over so-called 'quota-hopping' in which foreign skippers, many of them from Spain, buy UK quotas advertised in the British trade press. It is estimated that 160 'flag-of-convenience' boats take over 40 per cent of Britain's quota of hake and plaice. Rules have been introduced to require these boats to land 50 per cent of the catch in UK ports, have half the crew made up of British fishermen, or originate four trips from UK ports. Quota-hopping is only important, in Scottish terms, in the south-west.

Landings in Scotland – Peterhead is the biggest port in the UK – have remained stable in recent years with around 3000 vessels bringing ashore 500 000 tonnes of pelagic (occupying open water), demersal (living close to the sea bed) and shellfish species. In 1995, the pelagic catch, including herring and mackerel, was heavier, but worth just £26m, compared to the £181m for demersal species including cod, haddock and sole. Shellfish accounted for landings of 51 000 tonnes, worth £87m. In the same year, foreign boats landed 48 000 tonnes of cod, haddock, whiting, herring and mackerel.

Some Shetland and north-east boats, following the example of French vessels operating out of Kinlochbervie, have been pursuing deep-sea species off the edge of the continental shelf as an alternative to declining traditional stocks. In 1994, foreign vessels landed more than 1300 tonnes of fish including blue ling, orange roughy, round-nosed grenadier and black scabbard, with UK vessels landing just 112 tonnes. There is no quota set for these fish and very little is known about their biology, although they are known to be very slow-growing and may live for thirty years before reproducing. As a result, the stocks are vulnerable and there is concern that the fishery is unregulated.

Shellfish are landed principally in Fraserburgh, Mallaig and Oban, with edible, velvet and shore crabs, lobsters, Dublin Bay prawns or scampi, and scallops forming the key species. Potting for crabs and lobsters continues along the west coast, with trawling for shellfish in the Clyde, the Minch and the Fladden ground east of Shetland and Orkney.

Unfortunately, an imbalance still remains between the health of the stocks and the fishing effort. Put simply, there are too many boats chasing too few fish and, according to the scientists, the fleet needs to be reduced by between 20 and 30 per cent by 2002. The EU agreed, in April 1997, to a 30 per cent reduction of the fleet, and days at sea, over five years. The deal means that EU countries will have to reduce their fishing effort for the most threatened stocks – including cod, herring, mackerel, plaice and sole – by 30 per cent, and for the over-fished stocks – haddock, saithe, hake – by 20 per cent. The fishing effort for all other stocks will be frozen at present levels until 2001.

One of the problems for those trying to regulate the industry is the speed of technological development, which makes it increasingly easy to locate and catch fish. Technical conservation methods include regulations on mesh sizes to allow younger fish to escape. Each EU state has to follow a programme to reduce the fleet capacity to an agreed level, and there has been a reduction in the total tonnage of vessels at sea of around 10 per cent since the start of the 1990s, most of it concentrated in the small, remote ports around the north-east coast – Lossiemouth and Buckie, for example – and in the west-coast ports. By comparison, the healthy fishing areas, Lerwick and Peterhead, have increased their fleets.

Times have changed for Scotland's fishermen. In the 1930s, around 50 000 British fishermen brought home huge quantities of cod, haddock and plaice; today, less than half that number, in 8000 vessels, struggle to fill their nets. Further restructuring – most of the reduction has been brought about by decommissioning, with over 450 boats scrapped in the UK between 1993 and 1996 – will raise difficult questions. Should the small, vulnerable ports be helped, or should market factors be allowed to decide who goes?

The key North Sea stocks under threat are plaice, cod and herring. The business of reducing the effort directed at these species will be handled by the industry, scientists and politicians. The environmental movement should also have a role to play. Unfortunately, fishing is one of the areas in which least progress has been made in the past decade towards a meeting of minds on sustainable development. Too many environmental groups characterise fishermen

Fragile Land

as careless pillagers of the sea, and too many fishermen insist there are enough fish for everyone and question the advice of the scientists. The issue of fish populations is complex, with huge natural variations between 'year classes' of fish often complicating estimations of a healthy stock. For example, it is possible that one million haddock will be added to the stock one year, and 20 million the following year. All environmentalists, and most responsible fishermen, accept that over-fishing is going on.

Other concerns from the environment lobby include: the (mainly Danish) industrial fishery; dredging for scallops; the conflict between breeding seabirds and fisheries; 'black fish'; and seal numbers.

Fisheries inspectors estimated in 1997 that up to half the landings of North Sea cod and saithe, both endangered stocks, were illegal black fish. The practice has been going on for several years, and involves skippers landing over-quota fish at small ports for private sales agreed at sea. Lorries then pick up the fish and take them to processing factories or wholesale markets in, for example, Hull or Grimsby. Fishermen justify the black fishing by saying that quota limits are threatening their livelihood, while illegal fishing undoubtedly threatens to undo the benefit of con-servation methods. The problem is being tackled by Government plans to make all vessels land in designated ports. In the near future, it might also be possible to track trawlers at sea by satellite.

The differences of opinion between fishermen and environmentalists are often reduced to the modern phenomenon of clashing urban and rural cultures. Or, as one fishermen's leader put it to me: 'The school-educated fishermen against the university-educated tree-huggers'. Even the Worldwide Fund for Nature, which is more establishment than most non-governmental organisations (NGOs), lowered its standards to produce a series of adverts shown on MTV, the music channel watched by 70 million young people. One ad showed two fishermen carrying nets and running towards the centre of a beach in a competition to dive on top of a single fish – the 'last fish' – as it flapped

110

in the water. The image was catchy, unconstructive and inaccurate.

Although fishing is not party-political, UK interests suffered under the last years of the Conservative government due to its anti-European posture, and some fishermen's leaders believe that a period of Labour government will be good for business. Meanwhile, there are signs of a more enlightened outlook developing within the industry. The Scottish Fishermen's Federation decided in 1997, at last, that it should publish an environmental policy statement. The document is not wholly 'green', but recognises 'that the future of the fishing industry depends upon the sustainable harvesting of the seas without causing irrevocable damage to the marine environment'. The federation states that fishing policy should be evaluated within an environmental and socio-economic context, and suggests – seals again? – that regulation should involve all non-fishing impacts on the marine environment, including predator populations, marine pollution and fish farming.

Impacts of Fisheries
More than 40 per cent of Atlantic 'finfish' stocks are said to be seriously over-exploited, 15 per cent over-exploited and 36 per cent fully exploited. There have also been local collapses in shellfish stocks, and the cockle fishery on the Solway Firth has been closed since 1992 following intensive prosecution. Crab fisheries are not thought to be over-exploited, and the current level of fishing for prawns is thought to be sustainable. Lobster stocks are believed to have declined by up to 40 per cent in the Western Isles since the early 1980s, and stock enhancement projects are being proposed with juvenile lobsters being released on the sea-bed. Fishing also has an effect on birdlife and marine mammals, such as seals and harbour porpoises which are caught in nets.

Industrial Fishing
For around forty years, industrial fishing has exploited fish stocks not for human consumption, but for the raw

materials of fish meal and oil which are used to produce products including oil for margarine and cooking fat, meal for farmed fish, meal for poultry and pigs, and fuel for power stations in Denmark. Some of these uses are likely to strike the man in the street as unjustifiable, if not bizarre.

The world's largest producers of fish oil are Peru and Chile, although the North Sea fishery accounts for over one million tonnes of fish, which is more than half of all fish landed each year. Denmark is the world's sixth largest producer of oil and meal, and by the far the largest in Europe. In 1995, it produced 106 000 tonnes of fish oil and 374 000 tonnes of fish meal, 90 per cent of which came from the industrial fisheries in the North Sea. Since 1985, the fishery has been based on sand eels, pout and sprat, which are a vital link in the marine food chain, and are preyed upon by fish, sea birds, seals, porpoises and even whales. Conservationists believe the collapse of some fish stocks, and bad breeding seasons for some birds in the Northern Isles, can be attributed to industrial fishing. Sand eels are a major part of the diet of guillemots, puffins, fulmars and terns.

Although the UK is not a significant industrial fisher, it is the largest consumer of fish meal and oil in Europe and imports it from Peru, Iceland, Denmark, Norway and Chile. Around 70 per cent of the oil goes into domestic and commercial margarines and solid cooking fats.

The key competitors for fish oil – which seem more acceptable – are vegetable oils, including rapeseed, sunflower, palm and soyabean oils. But fish oil is cheaper. Margarines using fish oil are found in popular supermarkets in Scotland, and commercial cooking fats containing fish oil are used by biscuit manufacturers.

In 1994, the UK animal feed industry accounted for 65 per cent of fish meal imports, with the remainder used in aquaculture and pet foods. The industry has now moved into the Wee Bankie area, 40km off the Forth Estuary (outside the limit of territorial waters), in a key area for seabirds. In 1993, 99 000 tonnes were taken at the site and in 1995, 32 000 tonnes. In March 1997, environment and fisheries ministers from the North Sea states agreed to ban the fishery in some ecologically sensitive areas, but not the Wee Bankie.

Under the terms of the 'precautionary principle' enshrined by Agenda 21 in Rio, the industrial fishery is undoubtedly unsustainable at present. This is particularly so because, unlike the other stocks, there is no limit set on the catch. Shetland has the only significant industrial fishery in Scotland, and is proving that sand eel fishing can be sustainable. Its annual 3000-tonne fishery is arguably one of the best regulated in the world.

Oil and Gas

The oil industry is of major importance to the Scottish economy, employing around 64 000 people. The serious business began after the discovery of the Forties Field in 1970, and other key fields in the East Shetland Basin and the Central North Sea. Oil is piped ashore at Nigg on the Moray Firth and Cruden Bay, Aberdeen, and is transferred by onshore pipeline to Grangemouth. Pipelines also take oil to Sullom Voe on Shetland, Europe's biggest oil terminal, and to Flotta on Orkney. There are around 150 oil and gas production platforms producing 100 million tonnes of oil and 40 billion cubic metres of gas every year. Oil production rose in the Scottish North Sea from 83 000 tonnes in 1991 to 117 000 tonnes in 1995, although the main exploration activity has since switched to the west of Shetland, where BP's Foinaven field was the first to produce oil, and to the Atlantic Frontier around Rockall.

The onshore development of the industry, including the laying of pipelines, has been handled well, but there remain concerns about the effects of oil pollution around platforms, oil spillage from tankers and service vessels, and the business of decommissioning the North Sea structures, such as the Brent Spar, at the end of their working lives.

From the first stages of oil exploration, the industry has affected the seabed, with the discharge of drilling fluids, muds, and oil itself. It is estimated that in 1994, around 100 000 tonnes of chemical substances were discharged into the North Sea. The muds used to cool and lubricate the drill bit contain heavy metals and organic materials which are released into the sea; oil-based muds have been banned by Norway and the Netherlands, but are still used in the UK sector.

The issue of decommissioning was raised in spectacular fashion by the case of the 14500-tonne *Brent Spar*, a redundant oil storage facility which became a *cause célèbre* in 1995 when Shell announced that it planned to dump it in the deep waters of the Atlantic. Greenpeace responded with an astonishingly successful campaign in which activists occupied the platform at sea. The environment group won public support throughout Europe, particularly in Germany and Holland, and forced Shell to abandon its plans.

In strict scientific terms, dumping the Brent Spar at sea was not a bad choice, but the proposal raised more fundamental questions. Should the enormously wealthy oil industry be seen to abandon huge structures at the end of their useful lives? What signals would such actions give to consumers about modern lifestyles, and could they be excused under the precautionary principle? Should the sea be used as a litter bin?

Bathing Beaches

A 1990 survey of the coastline found that 78 per cent of estuarine waters and 84 per cent of coastal waters were of good quality. In 1997, 18 of Scotland's 23 designated bathing waters passed the European directive for standards based on 'faecal coliforms' – sewage, to you and me. However, the pass rate fails to tell the full story of the state of Scottish beaches, which are affected by 1000 authorised discharges, 800 of which are sewage effluent. The directive only identifies beaches which have a large number of bathers and a wide range of facilities. Neither condition applies to the majority of Scottish beaches.

The failures in 1997, all in the south-west, were Sandyhills (Dumfries and Galloway), Girvan, Ayr (South Beach), Irvine-Gailes, and Saltcoats/Ardrossan (South Beach). Girvan has the distinction of never having passed the test in nine years.

However, the prospects for swimming in Girvan were improved by the completion in 1997 of a new sewage system. Anyone planning a beach holiday may be interested to know that the six beaches which have passed every year are Cullen (Morayshire), Aberdeen (North Beach), Montrose, St Andrews (West Sands), Gullane (East Lothian) and Pease Bay in the Borders.

However, some of these have simply met the mandatory standard, and have not complied with the more desirable 'guideline' standard. Change is being driven by the European waste water directive which requires secondary, biological, treatment of all coastal outfalls serving more than 10 000 people, and all estuarine outfalls serving more than 2000 people, by 2005. The three Scottish water authorities expect to spend £1.6 billion to ensure compliance.

The Scottish Environment Protection Agency (SEPA) has been carrying out its own tests on other beaches not covered by the directive and in 1996, eleven of the fifty-six beaches tested failed the mandatory standard. These included Largo East, Kirkcaldy (Linktown), Kinghorn, Burntisland, Port Seton, Longniddry, Eyemouth, Lossiemouth East, Sandend and Newburgh.

All of us should bear in mind that improving beaches is not just about the action of the water authorities. It is a process which also involves individual action, such as taking litter home, and disposing of sanitary products and plastic items in bins, not toilets.

Aquaculture

Salmon farming began in the Highlands, in sheltered sea lochs, in the 1960s. The business remained relatively small through the 1970s as the problems of breeding, disease and cage design were unravelled. Most of the work was carried out by multinational companies such as Unilever and Booker McConnell, and the real expansion began from the end of the 1970s, with production rising from 1000 tonnes in 1980, to 70 000 tonnes in 1995.

Initially, enterprise agencies encouraged individuals to become involved, but today the industry is dominated by the big players. In 1996, SEPA regulated 108 companies producing just over 83 000 tonnes of fish from more than 250 sites. At the current rate of growth, the industry will be producing 132 000 tonnes by the year 2000. Since the early days it has become an important source of employment and has helped to revive ailing rural communities.

Salmon farming involves the use of freshwater sites where the young salmon are kept for around eighteen months until they become smolts and are ready to go to sea. The smolts are often flown by helicopter to the familiar cages at sea where they reach marketable size after about two years on a special diet.

The area of the salmon farm, effectively the seabed, is leased to fish farmers by the Crown Estate Commission (CEC), whose profits go to the Civil List which supports the royal family. The business grew rapidly, and was largely unregulated through the 1980s, with farms being approved in some of the most scenic sea lochs.

The CEC has now published guidelines on the location, spacing and appearance of cages, and can demand an environmental assessment for units over a certain size. In practice, this is rarely done; of the 1000 farms granted leases since 1988, only one required an environmental statement. Statutory responsibility for the industry is divided between the Scottish Office, the Crown, the Department of Transport, SEPA and planning authorities. SEPA has responsibility for preventing any damage to the marine environment, but, due to its current under-funding and the lack of staff available in the Highlands to visit marine farms, it is almost impossible for it to keep a close eye on the industry.

The key environmental issues are the use of chemicals to stop sea-lice infestations, the nutrient-rich feed, faeces, and escapees competing or breeding with wild fish. One chemical used for sea lice, dichlorvos, has been restricted because of its effects on marine plankton and invertebrates, but no ideal substitute has been found. Research is continuing, meantime, on a vaccine for sea lice which could take ten years to bring to the market, and some marine scientists are concerned that the alternatives to dichlorvos may be even more damaging. Cypermethrin is toxic to crustacea, but is meant to break down after one hour, and does not have negative effects on lobsters and mussels. Azamethiphos is meant to take the place of dichlorvos, but sea lice are already gaining resistance to the latter, and could do so with the new treatment. The replacements may also be almost undetectable at sea, making compliance with usage limits difficult to police.

SEPA plans to monitor and control any effects, and will also set limits for nutrient input – nitrogen and phosphorus – from feed. The organic load discharged by salmon farms is about half a tonne per tonne of salmon grown, and research suggests that when sea cages are moved the seabed and the native fauna take around two years to recover. The extent of the impact, which is localised, depends to a large extent on the nature of the water and the degree of flushing and water exchange. Farms in open sea areas have much lower impacts.

Salmon and sea trout proprietors are also concerned at the impact of aquaculture on wild stocks and the possibility of disease being transmitted. Many anglers associate the collapse of the sea trout fisheries in Loch Maree and Loch Eilt with the spread of salmon farming. There are also concerns about interbreeding, and the loss of genetic purity in wild stocks which for hundreds of years have been returning to the same rivers. The industry has answered some criticisms in the past by separating year classes of fish, and by fallowing individual sites, and sometimes whole systems, to allow the seabed to recover.

Like the fishermen of Orkney and Shetland, salmon farmers would like to see the number of seals reduced, although they are allowed to shoot seals which are raiding their cages: the mammals take chunks out of salmon by biting them through the nets.

Overall, although the industry has been much-criticised, it is now very important in economic terms. By making salmon cheaper, it has also helped conservation and angling groups to buy-out the former commercial netting operations on the main Scottish rivers.

SEPA must consider the question of sustainability in relation to the salmon farming industry, and it may be that production limits have to be set. But I would rather see salmon cages in sea lochs – where they are sensibly sited – and see people living in remote communities, than have pristine scenery and a declining population. It is claimed that the industry, directly and indirectly, employs 6000 people, which is more than the Scottish coal or steel industries.

The smaller business of growing shellfish is concentrated in lochs and bays in the west Highlands, where 315 farms produced around 1250 tonnes in 1995. Seventy per cent of the total was mussels, with the remainder divided between oysters and scallops.

Shellfish feed on natural plankton and do not need artificial feeding or treatment with chemicals, therefore their production has a limited environmental impact, although waste can accumulate on the seabed.

North Sea Pollution

Discharges from industry, sewage and agriculture are declining in Scotland. However, the UK is involved in an ongoing debate with the EU over the banning of all sewage sludge dumping at sea in 1998. The UK position is that it is not necessary to build extremely expensive plants to remove nutrients from sewage before the effluent is released to the North Sea. Countries on the other side of the sea, where there are problems related to over-enrichment, take a different view.

According to Scottish scientists, there is no evidence of a coastal eutrophication problem in the Forth and Clyde estuaries. At the Garroch Head dump site in the Firth of Clyde, only 10 per cent of the contaminants are thought to remain in a 3 sq km area.

However, there is general agreement about the phasing out of all polychlorinated biphenyls (PCBs) – oily substances used as insulators in transformers and other electrical equipment – by the end of 1999. The North Sea Conference of nations also aims to halve the amount of potentially harmful substances like mercury and cadmium reaching the sea from the atmosphere.

Such sensitivies have come about after decades of dumping at sea. More than one million tonnes of conventional and chemical weapons have been deposited in the sea around the UK since the last war, particularly in the seven-mile long Beaufort's Dyke trench in the North Channel between Scotland and Ireland. In 1995, more than 4500 incendiary sticks were washed up on the shores of the Clyde estuary when a gas pipeline was laid on the seabed between the two countries. However, a seabed survey found that levels of heavy metals were similar to safe levels elsewhere.

Most sea pollution comes from the land, and there are concerns about enhanced levels of organochlorine pesticides, PCBs, heavy metals, nitrogen and phosphorus. The central and southern sectors of the North Sea receive the majority of the river-borne inputs.

When a new chemical is marketed in the UK, the manufacturer or importer must provide detailed information on its properties. The Health and Safety Executive (HSE) and the Department of the Environment (DoE) then evaluate its potential to harm man or the natural environment, and share the information with other EU countries. However, there is little information in existence about the environmental effects of the 110 000 different chemicals marketed in the EU before new legislation tightened the procedure in 1981.

It often takes major incidents to precipitate new laws on the use and transport of polluting materials. The wreck of the *Braer* oil tanker, for example, led to an inquiry by Lord Donaldson, who put forward more than 100 recommendations on issues including routing, the provision of salvage tugs and the identification of environmentally sensitive areas.

Due to the light nature of the oil it was carrying, and the storms which lashed Shetland at the time, the effect of the *Braer* on the marine ecology of south Shetland was limited, although crab and lobster fisheries were closed for between twelve and eighteen months, and the scallop fishery was shut for two years.

A more recent concern of environmental groups is the use of hormone-disrupting chemicals found in detergents and industrial processes, which can have a feminising effect on fish. There is uncertainty at present about what constitutes safe levels of these chemicals, but according to Friends of the Earth (FoE), these 'gender bender' chemicals, in shampoos, hair colourings, shaving products and condoms, have been implicated in the rise in human breast and testicular cancer, and the decline in the average sperm count.

Litter from ships is another key marine issue. Around fifty species of birds, as well as fish, turtles and marine mammals, are known to eat plastic, but there is currently no requirement for ports and harbour authorities to have waste collection facilities. Large amounts of sea-borne litter build up on Scottish beaches from the Clyde to Shetland.

General Outlook

Conservation protection is uneven around the Scottish coast, and there is no national framework for management of the coast and the marine environment. Many important areas are not recognised under European wildlife law, and a limited number of areas – less than twenty – have been proposed as candidate marine Special Areas of Conservation under the Habitats directive. There is little sense in managing the marine environment on a sectoral basis. A draft planning guideline in 1997 recognised the need for local authorities to develop a co-ordinated approach known as the Coastal Zone Management (CZM) system. Support for proper coastal management has been growing since the 1980s, and is reflected in the establishment of SNH's Focus on Firths initiative, and the Government's Scottish Coastal Forum, which brings together interested groups.

SNH itself has a remit for the marine environment, but has been slow to tackle the issues. It failed to deliver Scotland's first Marine Nature Reserve at Loch Sween in Argyll. The proposal was handled clumsily and the local population was strongly opposed to what it saw as interference by an outside body. SNH has now decided the designation is less than ideal and has no plans to promote the system else-where.

The last Conservative government said that the pressures on UK coastal waters did not merit a large-scale, integrated management regime, but the evidence suggests otherwise.

6

·

The Air We Breathe

·

... this most excellent canopy, the air, ... why, it appears no other thing to me but a foul and pestilent congregation of vapours!

Shakespeare, *Hamlet*

Better, but Still Life-threatening

·

FIFTY YEARS AGO, fresh air was unknown to millions of Scots. Today, by world standards, our air quality is very good. Even when we suffer pollution, the concentration of unseen contaminants is lower than in other parts of the UK and Europe, and much lower than the world's most polluted conurbations. Glasgow's air is unfailingly sweet when compared with that of Athens, Mexico City and Bangkok. But we

122

have no room for complacency. Modern concerns about relatively low levels of pollution are sharpened by a much greater scientific understanding. Even in smog-free Scotland, pollution is thought to claim an extra 2000 lives each year, and the levels of pollutants which we regard as acceptable might be criticised in Norway and Sweden.

The unhappy experience of breathing dirty air was recognised as long ago as the thirteenth century, when a law was passed in London to prohibit the use of coal because of its effect on health. Nearly 400 years later, in 1648, Londoners asked Parliament to ban the import of coals from Newcastle, and as recently as the 1940s and 1950s, buses had to be guided on foot through the smog. In December 1952, a six-day London smog caused 4000 extra deaths in one week and precipitated a public outcry which resulted in the Clean Air Act of 1956. People were obliged to restructure their lives, and did so, as coal fires were put out in tens of thousands of homes. The creation of smokeless zones increased the average visibility in London on a winter day from one-and-a-half to four miles. Another Act of Parliament followed in 1968, and the two pieces of legislation probably led to the single biggest environmental improvement ever witnessed in the UK.

Now, our air looks good but we are learning about the potential dangers of minute 'trace' quantities of pollutants such as lead and benzene. We know, for example, that a few minutes of exposure to high levels of carbon monoxide will cause headaches and nausea, while constant exposure to benzene at concentrations of five parts per billion (the equivalent of five seconds in thirty-two years!) could be fatal.

The more you look at the air, the more you find. Canada, the USA and the Scandinavian nations are leading the way in developing policies to deal with indoor pollution caused by materials including insulation, chipboard, paint and cigarette smoke, which have been taken for granted for many years.

Pollutants also occur naturally. The ideal mix of oxygen, nitrogen and water vapour is tainted by volcanoes and forest fires. In most circumstances these are not especially hazardous, but since the discovery of fossil fuels we have radically altered the natural concentrations of pollutants. The industrial revolution brought huge outpourings of smoke from countless chimney stacks.

In the 1990s, winter sulphur dioxide levels in Prague were higher than the World Health Organisation (WHO) guidelines, and doctors advised pregnant women to leave the city. Even in Paris, where an excellent public transport system fails to prevent chronic traffic congestion, around 400 premature deaths are caused each year by pollution, and public transport is free when the air is particularly foul.

In Scotland, hospital admissions for asthma, which can be triggered by traffic pollution, have more than doubled in the last decade. More than 300 000 Scots, one third of them children, suffer from the respiratory disease. The acid deposition caused by air pollution also damages buildings, animals and plants.

In summary, we have two pressing problems: localised pollution caused by traffic in cities, and airborne pollution in the countryside caused by emissions from industry, power generation and traffic. In Glasgow, Edinburgh, Dundee and Aberdeen, human health suffers; in the Trossachs, the Cairngorms and Dumfries and Galloway, the heather moorland and the freshwater lochs and streams suffer.

The acceptable levels of sulphur dioxide, smoke, nitrogen dioxide, ozone and lead in the atmosphere are regulated by EU directives, while WHO issues its own separate guidelines. In March 1997, the Conservative government promised national policies to achieve broad reductions in pollutants, and gave local authorities the task of monitoring local air quality and designating air quality management areas. It also charged them with producing a plan of action, but did not provide additional resources for the task, or couple the whole plan to an integrated transport policy. The Labour Government has recognised, at least, that a national transport strategy is central to air quality. It should also act to improve the monitoring of air pollution in the UK, which is woefully inadequate when compared with other EU nations. There are just eight major air quality monitoring sites: Edinburgh; Aberdeen; Glasgow (three sites); Bush, south of Edinburgh; Eskdalemuir in the Borders; and Strath Vaich, west of Dingwall. One site in Glasgow is in a back street, while Edinburgh's monitoring is carried out in Princes Street Gardens. Neither location seems to comply with the EU directive which says pollution should be tested in 'canyon streets' containing heavy traffic.

Finally, we should remember that the pollution which we produce in our everyday lives, by switching on lights and driving cars, and taking buses, is also significant on a global scale. We export acid clouds to Scandinavia, and our emissions of carbon dioxide (CO_2) contribute to global warming. Global warming, remote though it seems, poses the greatest single environmental challenge to the world.

Anthropogenic Pollution

Carbon Dioxide (CO_2)

CO_2 is our main contribution to climate change. The gas is released by burning coal, oil and gas, particularly in vehicles, power stations and heavy industry. It is also freed to the atmosphere when trees and crops are burned and is released when living things, including us, breathe, die and decay.

Life depends on a natural global warming, largely caused by water vapour in the atmosphere, which keeps the world 308C warmer than it would be without its 'thick' atmosphere. The warming is caused when the sun's rays are trapped by gases instead of being reflected from the earth's surface back into space.

The global warming which faces us today is caused by the man-made pollutants which we have been adding to the natural cocktail for hundreds of years, and at a greatly accelerated rate following widespread industrialisation.

In Scotland, 31 per cent of CO_2 emissions come from transport, 25 per cent from industry, 15 per cent from services, and 22 per cent from domestic sources. And in 1989, we produced 44 million tonnes of the stuff.

Methane, another warming gas, is released when organic matter decays in marshes and wetlands, and is also produced in animal wastes.

Sulphur Dioxide (SO_2)

SO_2 is released by burning coal, smokeless fuel and oil, and forms weak sulphuric acid when dissolved in rain. Around 70 per cent of Scottish emissions come from power generation. In combination with smoke, it can cause temporary breathing problems. It is one of the two principal causes of acid rain and a major cause of building erosion. Although we have adopted the expression acid rain, over 50 per cent of SO_2 pollution is in the form of 'dry deposition', which takes place without rain or snow, near the point of emission. The Longannet coal-fired power station on the Forth estuary is a significant point of emission.

SO_2 is the most highly monitored gas in Scotland, with hourly concentrations recorded in Strath Vaich and Edinburgh, and daily concentrations taken at another forty sites to assess compliance with an EU directive. Pollution levels have been declining over the last twenty years, largely due to the decline of heavy industry. The total UK emissions sank from 6.4 million tonnes in 1970 to 4 million tonnes in 1980, and the official records between 1993 and 1995 suggested the health guidelines were not exceeded.

However, an official air quality panel, which set a new standard of 100 parts per billion (ppb), has questioned these results. It found that when they were converted from daily averages into 15-minute averages, the standard was exceeded in 33 per cent of the sites. When continuous monitoring was carried out in Edinburgh, the standard was broken on thirty occasions in 1994.

The EU directive on 'large combustion plants' requires emissions from power stations and incinerators to be reduced to 60 per cent of 1980 levels by the year 2003. Scotland's largest urban areas typically experience concentrations of the gas ranging from 10 to 20ppb, with occasional peaks of 400–750ppb.

Nitrogen Dioxide (NO_2)

Nitric oxide and nitrogen dioxide are gases formed during combustion from the nitrogen found in fossil fuels, and by the oxidation of nitrogen in the air. High concentrations can affect the respiratory system, inducing asthma and bronchitis attacks, and reduce plant growth. The gas is also responsible for one third of Scotland's acid rain.

Around half of the nitrous oxide (NO_x) emissions in the UK and Europe come from road transport, with a further 30 per cent from power stations. Britain agreed in 1984 to a 30 per cent cut in 1980 levels by the end of the 1990s, but there has been substantial traffic growth since then, and the target is unlikely to be achieved.

The three sites where hourly concentrations are monitored show a slight decrease in this type of pollution since 1980, although in 1994 none of the sites exceeded EU standards. However, when kerbside sites were tested in Glasgow, Edinburgh, Aberdeen, Edinburgh, Musselburgh, Dalkeith and Motherwell, the measurements indicated these sites would exceed health guidelines.

The natural background level is between 1 and 4ppb, while European urban areas range from 10 to 50ppb in bad conditions. Scottish statistics, up to 1993, show an annual average of 27ppb in Glasgow and Edinburgh. The EU directive sets a limit of 104ppb over a one-hour period, and an annual average of 21–26ppb.

Lead

Lead is the main heavy metal found in the air. It comes from leaded petrol, coal burning and metal works. Concentrations have declined to less than one third of the 1984 levels following the promotion of unleaded petrol. Lead can accumulate in the environment and in the body, and it is particularly harmful to children. It is known to damage the nervous system and the kidneys, and children may also suffer a loss of IQ and behavioural problems. However, it is now unlikely that EU and WHO levels will be exceeded. Lead is monitored in Glasgow and Motherwell, where levels have decreased by around 50 per cent since 1992–3.

Ozone (O_3)

Tropospheric ozone is a cocktail of gases, including oxides of nitrogen and volatile organic compounds (VOCs), formed through chemical reactions caused by sunlight. The troposphere is the lowest layer of the atmosphere, and the ozone found at ground level is separate from the stratospheric ozone in the upper atmosphere which protects life from the effects of ultraviolet radiation. High concentrations of low-level ozone irritate the eyes, nose and throat, cause headaches, nausea and inflammation of the lungs, and can also reduce crop yields, damage natural vegetation and contribute to acid rain.

Concentrations, monitored at just three rural sites and one urban site, are higher in the countryside because other forms of urban pollution have the effect of removing ozone from the air. Ozone concentrations are increasing and, in hot weather, levels in London often exceed WHO limits.

According to Friends of the Earth (FoE), the official statistics conceal the existence of most summertime ozone smogs because they describe ozone levels as 'good' unless hourly readings are above 90ppb. However, the government panel on air quality recommended a new standard of 50ppb averaged over eight hours. Background levels are usually below 15ppb, while the annual average for Strath Vaich in the Highlands is around 34ppb, compared to 13ppb in Edinburgh.

Carbon Monoxide (CO)

This colourless, odourless gas is formed by the incomplete combustion of petrol, and is measured continuously in Edinburgh and Glasgow, where levels are considerably lower than the proposed UK standard. More than 80 per cent of all emissions come from car exhausts, and exposure can interfere with the way blood transports oxygen, cause drowsiness, slow reflexes and impair mental and physical alertness. Background levels are around 0.01–0.2ppm, and the most recent average for Glasgow and Edinburgh was 0.6ppm, compared to a UK standard of 10ppm over an eight-hour average.

Particulate Matter (PM10)

Particulates are now thought to be one of the most damaging pollutants released by vehicles. They are produced by diesel engines, with 25 per cent of emissions coming from transport, and most of the rest from power stations, domestic coal use and incinerators. The smaller particles can stay airborne for ten days. They may cause respiratory and heart problems, have been linked to cancer and are thought to be responsible for 460 000 excess deaths worldwide every year. The World Health Organisation decided there was no safe level of particulates.

PM10 – the figure refers to the diameter of the particle – is most commonly used as an indicator of particulate pollution and is measured at just one site in Edinburgh. Particulate matter was responsible, with SO_2, for the famous city smogs of the 1950s. One study suggested that in 1994, the new guideline for the pollutant – 50 microgrammes per cubic metre – was exceeded in Edinburgh on thirty occasions.

Benzene, 1,3-Butadiene and VOCs

VOCs are organic compounds of carbon and exist as vapour. They come from the evaporation and use of petrol and diesel, and also natural gas leakage. Benzene is used in unleaded petrol and is also created in exhaust gases. Around 37 per cent of emissions come from road transport, 44 per cent from industrial solvents, and a significant percentage from cigarette smoke. VOCs are known toxins and can cause leukaemia. 1,3-Butadiene is a suspected carcinogen.

The UK has an average of 5ppb for benzene. It has been measured in Edinburgh since 1994 and levels are well below the health standards.

Ammonia (NH3)

Ammonia in the air is mainly the result of animal wastes produced from intensive rearing of pigs and other livestock. The emissions have increased in the past thirty years in the UK, although Scotland is a relatively small producer. However, it receives ammonia in rain from England, Belgium, the Netherlands and Denmark. Data on emissions and its effects are scarce, but it has been linked to the loss of heather and the spread of rough grass in some upland areas.

Radon

Radon is a naturally occurring radioactive gas produced from the decay of uranium, and is found in small quantities everywhere, but especially in areas of granite rock. It disperses quickly in the open, but can accumulate inside buildings. Exposure to radon can damage the lungs and increase the risk of lung cancer. It is thought that as many as one in twenty lung cancers might be caused by exposure to natural radon in homes.

Recommended levels indoors were halved in 1990 to 200 becquerels per cubic metre. Radon and thoron (another form of the gas) are responsible for 51 per cent of the average annual dose of radiation in the UK population.

The Issues
.

Global Warming

Yes, global warming is a Scottish issue. It may mean we can start growing grapes out of doors, but it could also bring more storms, more rain, less snow and the loss of plant and animal species.

If warming continues at the rate predicted by the 1996 UK Climate Change Impacts Review Group, average summer temperatures will be nearly 28 C higher in Scotland by 2050, with winters 3–48 C warmer than at present. Since the end of the last Ice Age, around 10 000 years ago, average temperatures have risen by only 28 C. The same rise is now being predicted in less than sixty years, and would mean a global mean temperature in the next century higher than at any time for 150 000 years.

The weather would then be too warm to support many species of Alpine plants and flowers on Ben Lawers, not to mention the Scottish skiing industry. And the range of the ptarmigan, snow bunting and mountain hare would be drastically reduced as species migrated northwards and to higher altitudes. Native species at the southern limit of their range would disappear, sea level rises of around 6cm a decade would lead to the loss of land in the Forth and Clyde estuaries, and there would be an increased risk of forest fires.

All the signs indicate that warming is already with us. In 1995, total carbon emissions reached six billion tonnes, an increase of 113 million tonnes since 1990. Carbon dioxide concentrations currently stand at around 360 parts per million, around 30 per cent higher than they have been in the past 160 000 years. UK emissions account for around 3 per cent of the world total.

It may be no coincidence that our spring weather arrived a week earlier in 1996 than it did in 1970, that UK studies have detected increased plant growth and photosynthesis, that work in the USA suggests the growing season is sixteen days longer than it was in the 1980s, and that January 1997 was the driest January in England for 200 years. Worldwide, there has been a series of record-breaking weather statistics since the early 1980s, when 33 000 animals died of thirst in one South African game reserve, the delta of the Okavango River in Botswana shrank by one third, and bush fires damaged 350 000 hectares in Australia. Ten of the hottest years in the 340 years of record-keeping have occurred since 1945, with three of the five warmest years occurring in 1989, 1990 and 1995. Six of the wettest years in the 135-year record of Scottish rainfall have occurred in the Highlands since 1983.

Since 1990, the world has also witnessed more than its normal share of natural disasters, with tornadoes, famines, forest fires, floods, water shortages and grasshopper plagues leading, among other things, to the collapse of some insurance companies.

Not surprisingly, these companies are now strong supporters of concerted action on global warming, and are cutting back on business in tropical storm areas.

Storms at sea are said to be growing fiercer and more

frequent, while there has been a 30 per cent drop in rainfall in the deserts of central Africa in the past two or three decades. The droughts which resulted caused hundreds of thousands of deaths and led to the 1985 Live Aid concert in London.

In 1997, fishermen reported early sightings of basking sharks, leatherback turtles and other exotic species in British waters, and in the USA, a study of migratory song-birds in New York State found that many species were arriv-ing significantly earlier than they did in 1900. Meanwhile, insect species are said to be colonising new ground in Europe, moving northwards at a rate of more than 2km a year.

Compelling evidence is also found in one the coldest parts of the planet – the Arctic – which has close and im-portant associations with Scotland. No less than 15 per cent of the world's bird species breed in the Arctic and depend on its short summer to raise their young. Many of them winter in Scotland.

In northern Eurasia, Alaska and Canada, snow cover was reduced by 10 per cent in the twenty years to 1992, with winter temperatures rising by up to 48 C between 1961 and 1990. The 1990 Arctic spring was the earliest in forty years, and one major reindeer herd found it was unable to reach its feeding grounds in time to benefit from the maximum plant growth. Elsewhere in the Arctic, warming weather has been bad for northern animals, and has opened the way for commoner species; Arctic fox numbers in the Fenno-scandian mountains are falling as red fox numbers increase. Conservationists are also concerned about the fate of wal-ruses, narwhals, musk ox, lemmings and bowhead whales, which are only found in the Arctic and are finely adapted to suit its climate.

As the natural treeline pushes north towards the Arctic Ocean, the area of permafrost and tundra for Arctic plants and animals is reducing. It is thought that some polar bears are experiencing difficulty in finding deep snowdrifts in which to shelter their cubs. Ringed seals also rear their young in snow caves, and warming weather makes the caves collapse, exposing the young to bitter cold. Ringed seals are the main prey of the polar bear, and any effect on them could threaten one of the world's most spectacular predators.

In Chapters 8 and 9, the means of reducing CO_2 emissions are considered in detail, but in broad terms it is fair to say that wholesale changes in behaviour are required at all levels in society, from the individual to the transnational company. Global warming is caused by unsustainable economies and lifestyles. Solving the problem requires international co-operation on an unprecedented scale, the development of new power supplies, new transport systems and hugely increased energy efficiency at home and in the workplace.

The scale of the task was recognised on the world stage in 1992, when the nations at the Earth Summit in Rio de Janeiro signed a convention on climate change which required CO_2 emissions to be stabilised at 1990 levels by the year 2005. Fortuitously, the UK is likely to meet the target because of the decline in heavy industry and the switch from coal to gas-fired power generation, although the contribution from traffic is still rising.

At the 'Rio-plus-Five' conference in New York in June 1997, the UK called on industrialised countries to aim for a 20 per cent reduction by the year 2010, but the proposal met with little support from the USA, the single biggest emitter of the gas. Robin Cook, the Foreign Secretary, remarked memorably at the time: 'Nobody is going to ask them [Americans] to embrace a life of poverty or a hair shirt, but there are other ways in which you can have a very good, advanced, enjoyable lifestyle, just as prosperous, just as rewarding, without having to drive everywhere in a very large car with a very large petrol consumption.' Unfortunately, many developed nations, including Canada, France, Italy and Japan, will not even meet the original target set in Rio. In 1996, emissions in the USA were 6 per cent higher than the 1990 baseline.

Acid Rain

This has been dealt with in Chapter 4, but merits some

further comment. The UK exports 75 per cent of its SO_2 emissions and 90 per cent of its NO_x output, yet the Trossachs and Dumfries and Galloway continue to suffer from serious acidification because of their high annual rainfall.

More than fifty sites of special scientific interest (SSSIs) are said to be affected by acid deposition which is damaging terrestrial and aquatic plantlife, insects and fish. The damage is done by increased acidity leading to the release of toxic aluminium from the soil.

A large proportion of the acidity suffered in Scotland can be deposited in just one or two falls of rain or snow. In fact, acid rain is a misnomer in the Scottish hills, where snow and cloud are often more acidic than rainfall. Acid pollution can travel up to 1500km from the point of emission, and sooty particles from power plants in Poland have been found in the Cairngorms. Around 12 per cent of the acidic depositions in Norway come from the UK.

Britain is committed to reducing sulphur emissions from power plants to 60 per cent of 1980 levels by 2003, and sulphur from all sources by 30 per cent on 1980 levels by the year 2000. It is likely that these figures will be reached, but detailed studies suggest that the targets are not strict enough to allow the recovery of Scotland's acid waters.

The Ozone Layer

The response of the international community to the 'holes' in the ozone layer caused by the release of chlorofluorocarbons (CFCs) and other ozone-depleting substances is a global good news story. The developed and developing worlds agreed in Montreal, in 1987, to limit and phase out the chemicals used in refrigeration, aerosols, foam and air conditioning which were destroying the thin layer in the stratosphere (15–50km above the earth surface) which protects all life from harmful ultraviolet (UV) radiation. Action was agreed after scientists concluded that further loss of ozone – a large hole had already opened up over the Antarctic – could have disastrous effects, with millions of extra skin cancers, reduced crop yields and damage to the marine plankton which forms the basis of life in the oceans.

By 1995, the agreements set in train had led to a 76 per cent reduction in the production of CFCs. However, CFCs and other ozone depletors are inert chemicals, thought to be benign when they were developed, which can last in the atmosphere for decades. Consequently, the ozone layer is not expected to recover fully, assuming all countries comply with the protocol, until around 2050.

The ozone eaters can take eight years to reach the stratosphere, where they break down under intense ultraviolet radiation at certain times of year, and release chlorine which reacts with ozone and converts it into ordinary oxygen. For the immediate future, an ozone hole will continue to appear over the Antarctic in early spring, while the layer thins in the northern hemisphere.

The United Nations Environment Programme calculated that for every 1 per cent loss of ozone, there would be an extra 50 000 skin cancers and 100 000 cases of blindness from cataracts.

The Montreal Protocol is now seen as the first evidence of successful, global environmental diplomacy. Although the decisions taken internationally affected some multinational chemical companies – notably ICI and DuPont, which initially played down the problem – substitute chemicals were found, and the issue did not involve a change in lifestyles. The two great examples of environmental diplomacy which followed the ozone agreement – the conventions on climate change and biodiversity signed five years later in Rio – have been much less successful because the issues are infinitely more complicated.

Ways Forward

Transport

Emissions of several key pollutants can be reduced by new vehicle technology. For example, on-board diagnostic systems may improve the emission performance of cars at a cost to the driver of less than £2 a year. Alternative fuels could also reduce air pollution. Compressed natural gas and liquid petroleum are both fossil fuels, but offer reductions in NO_x and particulate (PM10) emissions when compared to petrol and diesel.

Electric vehicles have no emissions, but carry heavy batteries and so displace the pollution to the point of power generation. It has been suggested that hydrogen produced from water, and used to create electricity in on-board fuel cells, may power the ultimate electric car. Hybrid cars are also being developed in which a small conventional engine runs constantly to generate the electricity needed to power electric motors. Bio-diesel from vegetable oils is a renewable fuel, but with higher NO_x and PM10 levels, while ethanol and methanol, which are cleaner than petrol and diesel, are used in Brazil.

However, the major changes required to improve air quality can only be brought about by integrated transport and planning policies which reduce the number of vehicles on the road.

Energy

In New York in June 1997, Tony Blair called for nothing less than a revolution in lifestyles to achieve significant cutbacks in greenhouse gases. The message was the right one, even if it was not obvious how he was going to deliver the promise.

One of the most obvious answers is improved energy efficiency in the home, the business and the workplace, and the promotion of electricity from renewable sources such as wind and wave power. Benign sources of power generation will become particularly important as Britain's nuclear power stations are decommissioned next century.

To help reduce CO_2 emissions from power generation, Gordon Brown, the Chancellor, reduced VAT on energy-saving products such as loft insulation, draught-proofing, hot-water-tank jackets and cavity wall insulation.

It is estimated that the right mix of measures could lead to a 50 per cent saving in energy use in Britain's housing stock, which is among the most inefficient in Europe. Only 14 per cent of Scottish homes have central heating, double glazing and wall insulation.

7

·

Agriculture

·

Farming, the only thing of which I know any thing, and Heaven above knows, but little do I understand even of that …

Robert Burns, *Letter to James Smith*, 1787

Common Sense and Good Food

·

IAN MILLER, of Jamesfield Farm, Newburgh, is an eloquent supporter of organic agriculture. He became increasingly frustrated during what he calls the 'mad 70s' as old trees were pulled down and wet areas were drained to make big fields. He resolved to 'go organic' when he discovered that soil fertility on his farm was slowly declining under a conventional, intensive regime. He reckoned that if he did not make the move, future generations would suffer because of his reliance on chemicals.

Miller says that before he converted he was a 'cosmetic' farmer who took delivery of bags of chemicals and followed the instructions on them. 'Eventually I couldn't understand why I was putting poison on food and poison on the land. I could see that the success of the farm no longer depended on the skills of the farmer. It was about the amount of chemicals you bought, and how you followed the instructions. I knew the eventual results would be disastrous, and future generations would have to face up to it. I thought, this can't be right, there must be something seriously wrong with our methods.'

His aim, on turning to organic production on his beef and vegetable farm, was to create a sustainable unit in which the farmyard manure from cattle, pigs and hens would provide the organic fertiliser he needed to enrich the soil.

He recalls hearing good advice from farmers of his father's generation, and laments the fact that 'a lot of knowledge has been taken to the grave'. 'Farmers ten years younger than me have had no experience of anything other than chemical solutions to pests and disease.' At the age of 52, Ian Miller is now relearning the relationship between his farm and the natural environment.

He dismisses as nonsense the mantra of conventional operators that organic farming is not capable of producing as much as conventional methods. He admits the changes he has introduced on the 300-acre unit since 1984 have not been easy, but says he is now producing more in all areas of the farm, with the exception of wheat. He is particularly pleased that he is able to grow supermarket-quality lettuces without any damage from pests. In the first few years of conversion – when yields dropped and no premium was paid – he encountered plagues of aphids and tried nettle juice to treat them. Now, a natural balance has been restored and there are more natural predators eating the natural pests. He sees ladybirds, songbirds, partridges and other wildlife which previously were missing.

'I feel that I can grow crops that are far better than they were with chemicals. I am criticised for saying that, but I don't care. We have had difficult times making the change, because there is not enough support, but neighbours are beginning to look at me now and say I am not crazy after all. We earn quite a premium for our produce. When I look back, I see I was guilty of removing old trees and hedges and making fields big in the mad 70s. We are now trying to put them back, and nobody has time to wait 100 years to see a tree grow. It all gets back to common sense.'

He adds that the previous government was totally uninterested in promoting organic farming, and hopes for more encouragement from Labour. 'Nearly ten years ago, I said farming would be changed by the public deciding they wanted food without chemicals. The long and the short of it is that the public have confidence in organic food being free of chemicals.' He also recalls that feeding animal parts to livestock was condemned by the organic movement long before the BSE crisis.

Duncan Shell, a 33-year-old farmer from Duns, Berwickshire, is another convert. In agricultural college in the mid-1980s, he took a radically different view from his lecturers, and disapproved of many of the conventional farming methods being taught. He disliked the reliance on chemicals, and produced a thesis on the potentially damaging effects of the continual use of nitrogen as a fertiliser. Even nitrogen, he discovered, had damaging effects in terms of plant composition and the chemistry of fresh waters.

For the past ten years, he has been putting his personal philosophy, which is also informed by his Christianity, into effect. Like Miller, he claims his mixed farm – 1000 ewes, 60 cows and nearly 300 acres of swedes and cereals – produces more than it did under conventional methods. And the produce attracts a premium in the marketplace.

Both men are members of the Scottish Organic Producers' Association (SOPA), whose fully organic members farm 17 500ha. The figure represents a very small percentage of the total farmland in Scotland, but interest is growing. In 1997, another 7000ha were in the process of conversion.

SOPA certifies and inspects organic producers in Scotland, in the same way that the Soil Association operates south, and north, of the Border. It was set up in 1988 and has seen a steady growth in interest, particularly through the summer of 1997, in the wake of the BSE crisis. It takes two years to convert to organic – to flush the system of chemicals – and over that period, and the following three years, the Organic Aid Scheme (a poor neighbour of the support systems in other EU states) provides financial help.

The organisation would like to see more money directed towards the scheme, which has been operating since 1994. It offers £70 per hectare for arable farms for the first two years and £25 in the fifth year, up to a maximum of 300ha. For hill sheep farming, the payments are much lower, reflecting the fact that it is already nearly organic.

Duncan Shell's most obvious initial problem was a reduction in grass yield. The species composition of pasture is to some extent determined by its management, and the composition had to be changed, with high levels of clover being encouraged to improve production. He now produces more grass than he did before, and the soil fertility has been improved with the use of rock phosphates, lime and seaweed. He no longer dips his sheep, and believes that organophosphate sheep dips, implicated in severe human illness, including ME or chronic fatigue syndrome, should be banned. 'I have a fairly philosophical attraction to what I am doing', he says. 'I am sure it has been a financial benefit to me, but it was certainly a risk to do it. It is a bit of a gamble, because there is a time lag in profits and it is not a thing you can easily jump into and make a quick killing. But it is the kind of farming which I much prefer, and I wouldn't ever want to return to conventional farming. I am a Christian and I believe I have stewardship responsibilities to the land and towards the livestock and the general population too. I do not believe in non-organic food, I am concerned about residues in food, and I don't think we know half there is to know about the dangers we are exposing ourselves to. As a small boy I used to ask my father not to use chemicals, because I always thought there must be another way.'

The farmers who have switched to that other way have found no difficulty in selling their produce. According to

Shell, there is a deficit in the quantity of organic vegetables being produced; wheat is undersupplied and oats are well supplied. Meanwhile, organic produce is said to be worth more than £10m a year to Sainsbury's, with sales doubling or trebling every year, although the retailer has to import 70 per cent of the total. At present, only 48 000ha, or 0.3 per cent of the land area, is under organic production in the UK, while other European states have converted great swathes to environment-friendly farming to supply every type of food outlet, from bakers to babyfood makers.

The Resource

·

Although intensification is the most obvious characteristic of modern agriculture, grazing systems still dominate in Scotland. Three quarters of Scotland is defined as agricultural land, though only 11 per cent is in arable production, 19 per cent is 'improved grassland' and 69 per cent is rough grazing. The vast majority, around 90 per cent, is regarded as a Less Favoured Area (LFA) by the European Union. This means it is unproductive land where farming has to be subsidised to prevent rural depopulation. Only 2 per cent of Scotland's agricultural land is in the highest land capability class. The widespread coverage of agriculture, however, means that a very large proportion of Scotland's wildlife is either found on, or uses, farm land of one type or another.

The nature, and the appearance, of farming today has been radically altered, first by the drive for self-sufficiency in Britain in the post-war period, and more recently by the much-maligned Common Agriculture Policy (CAP), which has been responsible for over-production and the infamous 'food mountains'. Under the current CAP, following reform in 1992, over-production has largely been restrained by a system of controls, including set-aside schemes for cereals, quotas for milk production, and quotas, headage limits and extensification payments for livestock.

At present, cattle numbers in Scotland stand at around 2.1 million, compared to 1.4 million during the Second

World War, and a peak of nearly 2.7 million in the 1970s. Sheep, however, are at record levels, having risen from 6.8 million in the 1940s to approximately 9.5 million today, replacing in many cases the traditional hardy cattle breeds of the Highlands. The predominance of sheep has reduced the cover of heather and encouraged the spread of mat-grass (*Nardus*), purple moor grass (*Molinia*), bog cotton (*Eriophorum*) and deer-grass (*Scirpus cespitosus*). Sheep, un-like the cattle they have replaced, are more selective grazers taking more of the flowering herbs, and leaving the coarser grasses.

High stock numbers, combined with excessive burning of the moor in some areas, reduces vegetation cover and can cause erosion of steep slopes and peaty soils. Soil erosion by water and wind can also occur on arable land in the east of Scotland, and is often associated with tillage for winter wheat – more than 400 incidents were recorded in 1984–5. Between the 1950s and the 1970s, there were substantial reductions in moorland and wetland through re-seeding and drainage. These activities are no longer grant-aided and the reclamation of rough grazing has almost ceased.

On lowland livestock farms, regular ploughing and re-seeding has replaced permanent pasture, and mown grass is usually made into silage, rather than hay. The cattle – mainly a few continental breeds – require housing for long periods, lambs are 'finished' on improved grassland, and lowland livestock are commonly fed on turnips or sage in the winter.

Farming has also become more specialised, with fewer mixed farms and the abandonment of traditional rotations. The area of wheat increased from 70 000ha in 1982 to 110 000ha in 1990, and the area of vegetables from 3500 ha to 11 500 ha. At the same time, small fields have been amal-gamated and more land has been cultivated.

Surprisingly, Scottish farms are larger, on average, than in any other region of the EU, but the amalgamation of fields, the end of arable cropping and the removal of hedges and trees have served to simplify the farming landscape, making it less interesting and less able to support wildlife. In many areas, dykes are falling into disrepair and are being replaced by fences. Hedges, where they survive, are often cut mechanically.

While mechanisation has increased – there are three times as many tractors today (over 60 000) as there were in 1946 – employment has fallen, with the farm labour force of 127 000 in 1921 now reduced to less than 28 000. Yields, however, have increased due to chemical applications, improved breeding and genetic manipulation. Farm buildings, meanwhile, have become much more prominent, with larger buildings for livestock and farm machinery. Grain silos and silage pits are common features, and the recent introduction of oilseed rape and black-bag silage have caused public concerns about the effects on the landscape and, in the case of oilseed rape, on human health. On a more positive note, planting of new woodland under a mechanism called the Farm Woodland Premium Scheme has met with approval from environmentalists.

Overall, the area of agricultural land in Scotland has declined in the past century, with most losses being caused by building, roads, and the conversion of grazing land to commercial forestry. In the 1980s, generous forestry subsidies resulted in large areas of rough grazing being converted to forestry in the Highlands, Strathclyde, Tayside and the Borders, and significant areas of arable and improved grassland going to forestry in Dumfries and Galloway.

Despite all the changes, Scottish farms can still be divided into four broad categories: hill livestock, lowland livestock and dairy, arable cropping and crofting.

Hill Livestock Farms

Over much of the uplands, farming is limited by climate and soil. Sheep farms in the hills, and sheep and beef cattle on the lower ground, predominate over 55 per cent of Scotland. These farms can define the character of upland scenery with their mixture of 'in-bye' land (the better ground around the farm/croft) and hill land, enclosures, steadings, sheep fanks and stone dykes.

A typical hill farm would have around 800 ewes on 1250ha of unenclosed hill land, and 40ha of in-bye. The output would be mainly store lambs which would be sent to the lowlands for fattening. A mixed livestock farm in an upland area would be made up of around 200ha of rough grazing, with 100ha of forage or improved pasture, and would support about 300 ewes and 30 suckler cows, which rear calves for beef production. It would produce store animals and finished stock, ready for the market. The farming method in this case is extensive, and relies on the semi-natural process of nutrient cycling and plant growth.

Lowland Livestock and Dairy Farms

The best lowland livestock and dairy farms are in Ayrshire, Dumfries and Galloway, the Borders, Orkney, Caithness and parts of Tayside and Grampian. A typical unit, creating a landscape of pastures, hedgerows and shelter belts, would cover around 200ha, including 65ha of arable crops and a small area of permanent grass. The remainder would be temporary rye-grass pasture, with housing for 200 cattle and grassland for 300 heifers or sheep. This system relies heavily on intensive methods to improve pasture for both grazing and silage.

Improved grassland covers 13 per cent of Scotland and is usually managed on a six-year rotation and treated with lime and fertiliser. The risks of pollution are considerable. Silage effluent, slurry and fertiliser run-off all affect water courses, as can detergents and disinfectants for dairy hygiene.

Arable Cropping

This is largely found in the eastern lowlands, where cereal farms can range in size from less than 100ha to over 250ha. The largest farms, with big, open fields, tend to be in the Lothians and the Borders. The main crops are winter wheat and barley, with smaller areas of 'break-crops' such as oil-seed rape or silage grass. The larger farms will often have some grassland for beef cattle and silage. There are also small arable farms found on the lower slopes of hills, particularly in the Grampians, where there are suckler cows and ewes, along with cereal and swedes.

Arable farming is highly mechanised and intensive. Crop

rotation is practised, but plants are also protected and promoted by inorganic fertilisers, pesticides, herbicides, fungicides and growth-regulators. The soils have often been well drained and are prepared for planting by ploughing and harrowing. Areas with sandy, permeable soils may have high-value crops irrigated in early summer.

Crofting

The croft lands, easily recognised by the patchwork of pasture, meadow, cultivated land and scattered walls and houses, cover one tenth of Scotland's agricultural land area, and are a highly characteristic feature of the environment of Shetland, the Outer Hebrides, Skye, Tiree, Wester Ross and Sutherland. There are around 17 000 crofts, most of which have a small area of in-bye or arable land – usually less than 10ha – and a share of more extensive common grazing on the hills, or on machair pasture, of up to several hundred hectares. Livestock rearing is the main activity, with store lamb production dominating. Significant cattle numbers – which are beneficial to the crofting habitat – remain only in a few areas such as Tiree, Uist and parts of Skye. Hay, silage and fodder can be grown on the in-bye land or the machair, but cropping is rare and crofting is invariably a part-time occupation supported by subsidies.

As a low intensity form of agriculture it has created and maintained important wildlife habitats and landscapes, has been less changed by developments in technology, and depends less on chemical treatments.

The Issues
.
An Overview

Agriculture has undergone a revolution in the past 100 years, and may be about to witness another one. The combined effects of mechanisation, chemical fertilisers and pesticides, indoor animal rearing, genetic engineering and new crop varieties has eliminated food shortages in Europe. To maintain the situation, the European Union pays subsidies, mainly headage and hectarage payments, to provide incentives and protect farm incomes, and it sets quotas to limit over-production. Overall, Scotland's farmland generates a gross annual output of £2.1 billion, with the help of £370m in EU and UK market support and direct subsidies of around £400m. In recent years it has also begun paying so-called agri-environment subsidies, in recognition of the fact that the intensification of farming has had damaging environmental effects. Occasionally, these environmental payments are an attractive extra for farmers, but they are not high enough to change the face of the countryside.

Now, with another round of world trade talks looming, wholesale change seems inevitable, though it is hard to say whether the environmental prognosis for the next revolution is good. The major change predicted is a general freeing of world trade conditions, which will make farmers compete across international boundaries on a level playing field. Consequently, it is almost certain that in the next ten to twenty years, the EU will move away from commodity-based support, and, in places like Scotland, place more emphasis on environmental and social subsidies.

Some of these changes are already being demanded by the public in order to satisfy ethical and environmental concerns. The BSE crisis, for example, not only damaged the profitability of beef farming, it shook customer confidence and stimulated interest in organic farming. The new look of the Scottish countryside must be economically viable and environmentally sustainable.

The Environmental Footprint of Modern Farming
Intensive agriculture has rendered the landscape simpler, less attractive to wildlife, and less pleasing to the eye. Polluting emissions to the air, land and water have increased, and chemicals have killed wildlife and infiltrated surface and ground waters with damaging effects.

Drainage has played a major part in changing the look of the land. Ditches have been ploughed, natural watercourses have been deepened, field drains have been buried below the soil, streams have been re-routed and straightened, and boggy areas have been drained and cultivated. This improved drainage has resulted in the more rapid run-off of rainfall, and an increased potential for flooding. The areas of production on lowland farms have been extended as close as possible to the field and farm boundaries, and where the boundaries are water, there is a chance of chemical sprays drifting directly on to watercourses. Since the 1950s, there have been detectable rises in the nutrient content of rivers in arable farming areas, leading to increased weed growth and increased concentrations of nitrates in groundwater.

Organophosphorus sheep dips have been linked to nervous system disorders in agricultural workers, and have leaked into water courses. The dips which are replacing them, synthetic pyrethroids, are toxic to aquatic insect life and numerous pollution incidents have been recorded.

Slurry (100 times more polluting than human sewage), silage and agricultural fuel oil are spilled into water courses, though the number of incidents has dropped since the introduction of new pollution regulations in 1991, and animal wastes are responsible for 25 per cent of the world's methane emissions, which add to global warming. Animal excrement also adds ammonia to the atmosphere, which oxidises to nitrate and can be carried far from the original source.

Soil is being blown and washed away as a result of bad management, burning of stubble, ploughing and re-seeding in autumn, the removal of hedgerows and the lack of maintenance of hedgerows.

Some of these issues, particularly the use of pesticides and herbicides, are being tackled by a Scottish National Heritage (SNH) initiative called TIBRE (targeted inputs for a better rural environment) which encourages farmers to become more careful with their sprays, a habit which delivers environmental benefits and saves money. The scheme includes the use of new spraying technology regulated by satellite, and encourages biological pest control systems such as living slug parasites which can be sprayed on a field and lie dormant until slugs appear.

Farming and Wildlife

There is a close relationship between the nature of farmland and its value for wildlife. In modern Scotland, most wildlife is found on the semi-natural rough grazing areas, and on land which is not intensively farmed.

Inevitably, most mammals, birds and plants have larger and more diverse populations outside intensively farmed land. All the features which have been deliberately removed from these areas – hedgerows, stone walls, farm woodlands, hedges and trees, stone walls, streams, ponds and wet places – have a high value for wildlife. Where farmland is characterised by a patchwork of small fields, woodlands and hedges, it will support a greater diversity of wildlife than the more familiar, open agricultural landscape. However, changes in the farm landscape, and in wildlife diversity, have often passed without notice because of the slow nature of their development.

The problems encountered by wildlife can be illustrated by the fact that around one quarter of the bird species in lowland farmland need to nest in woodland, while another fifth nest in hedges. Hedges are also a refuge for plants, insects and small mammals which were once much more widespread in a patchwork countryside, and are used by the predators which might exercise natural control of crop pests. Similarly, the richest habitats for flowering plants, and for insects such as butterflies and beetles, are the threatened ditches and streams.

Birds are particularly useful indicators of environmental change because they are near the top of the food chain and are easy to monitor. Farmland birds can be grouped according to the type of farmland on which they depend. For example, mixed lowland arable farms are important for birds in serious decline, such as the corn bunting, grey partridge and skylark, as well as wintering swans and geese. Better management of cereal field margins would help some species. The last government's Biodiversity Action Plan recommended urgent action to conserve seventeen species, and ten types of habitat.

One of Britain's rarest birds, the corncrake, depends for

its survival on the way that grassland is managed. Its breeding habitats worked well with traditional hay-making methods, but the expansion of silage has resulted in grass being cut earlier, and the young birds being destroyed. As a result, the corncrake, whose distinctive croaking cry was common in many parts of mainland Scotland forty years ago, is now largely confined to small areas in the Western Isles where crofters continue to practise traditional grassland management. In some of these areas, the Uists and Coll, for example, SNH and the RSPB are supporting corncrake-friendly management and the birds are recovering.

The corn bunting has suffered as a result of mechanisation since the early 1960s, and the chough has become restricted because its dominion was linked to invertebrates found during the out-wintering of cattle.

Grassland managed in a way which maintains wild flowers is rare, with some surviving in Perthshire, Dumfries and Galloway, the Borders, Skye and Lochaber. Herb-rich grasslands also survive in the machair areas of the Outer Hebrides, Coll and Tiree and parts of the Northern Isles.

Cultivated fields can support annual weeds which provide seeds and attract insects for bird life, but cornfield weeds, such as the corncockle, have become much rarer in the past sixty years due to the use of herbicides. Conversely, some weeds which are difficult to control with chemicals have become more abundant. Conservationists would like to see farmers avoid spraying, and even cultivate, field margins.

The effects of changing land use have also played a part in the extinction of eight species of butterfly since the early nineteenth century, and a restriction in the range of a further eleven species. The common blue and the meadow brown, linked to unimproved grassland, are still among the commoner Scottish butterflies, but their numbers have dropped. The birds-foot trefoil, the main food of the common blue, used to be sown by farmers.

Straw-burning after harvest, which is illegal in England, is another damaging practice north of the Border. It can significantly reduce the number of earthworms and other invertebrates in the soil. Worms are also less abundant in soils treated with inorganic fertilisers.

The Agri-Environment Programme

The EU and government programme which encourages environmentally positive farming includes the organic aid scheme, but more than 60 per cent of the financial assistance available goes to the ten Environmentally Sensitive Areas (ESAs) designated by the Scottish Office since 1987. They cover 20 per cent of the total agricultural area, including Breadalbane, Loch Lomond, the Central Borders, Shetland and the Argyll islands. Farmers have to volunteer for the scheme for an initial period of ten years, and are required to sign a management agreement based on a farm conservation plan. They will then receive payments for adopting methods which maintain or enhance the landscape, nature conservation or archaeological interest of an area.

Farmers have to be motivated to set up these schemes, and often claim they offer too little reward for too much effort. In 1995, the highest uptake was in Loch Lomond, where 56 per cent of the eligible farmers participated, but the lack of interest elsewhere meant that the quite modest ESA budget was not being spent.

When farmers sign up to an ESA agreement, they are obliged to have a whole-farm conservation audit to identify the most valuable conservation opportunities. There are eighteen options to choose from, ranging from the management of hay and silage fields to protect ground nesting birds, to the creation of wider and taller hedges.

The Countryside Premium Scheme, introduced in 1997, covers all the farmland outside the ESA schemes, and offers similar incentives for environmental improvement.

The agri-environment arrangements also include setaside. Between 1992 and 1997, 15 per cent of large arable farms had to be taken out of production every year to guarantee subsidies. The figure was cut to 5 per cent of the farm for 1997.

The final safety net for the farm environment is the

Agriculture Act of 1986 which requires agricultural policy and agriculture practice to take account of environmental objectives, and recognises that farmers have a dual role as producers of food and managers of the natural heritage.

At present, none of the support systems is tested, but the next CAP reform could require member states to prove that their agri-environment projects are having a positive impact. Such a condition would focus attention on Britain's international obligations under the Birds and Habitats directives to protect, for example, the corncrake, the chough and the barnacle goose, and habitats including machair, merse and arable field margins.

The Future of Subsidies

The World Trade Organisation talks, in 1999, will almost certainly make agriculture more competitive from early in the next millennium. The basic idea is that in a free market situation, food will be cheaper because it is produced without subsidies by the most efficient farmers on the best farmland.

Society wants farmers to supply cheap food produced to the highest standards, and at the same time wants economic activity maintained in rural areas. The Farm Assurance scheme, which provides detailed codes of conduct, exists to reassure consumers concerned over the way food is produced. But issues such as BSE, and the increasingly common outbreaks of *Salmonella* and *E. coli*, have led consumers to question modern practices in industrial farming. Changes on the world stage are likely to be accompanied by political pressure on ethical issues. The export ban imposed after the BSE crisis obviously damaged the £1.2 billion beef industry in Scotland, but the affair also resulted in a 15 per cent fall in domestic consumption.

The moves ahead will have important impacts on Scottish farming, much of which depends on subsidies. After 2003, it is likely that more payments will be made to fulfil social and environmental objectives, but they must be decoupled from production. In that situation, the survival of agriculture in the Less Favoured Areas which cover most of the country will depend on the development of new support systems with non-agricultural goals. In other words, keeping people in the countryside and enhancing the natural environment. Hill farmers are particularly vulnerable to change. Their subsidies are currently in excess of 100 per cent of the farm income, paying the income itself and covering the trading losses. These areas remain important, however, because they provide the sheep breeding stock on which all livestock is based. From an environmental point of view, the changes ahead may be appealing if they lead to upland areas becoming more wooded as grazing animals are removed from the most marginal land.

Those countries which want the removal of all support systems, including the USA, claim the move is necessary, partly to cope with the anticipated doubling of the world population by 2060, and the predicted food shortages. The opposite view – the 'fortress Europe' option – suggests that Europe should forget about exports and supply internal demand only to ensure a good standard of living.

At a local level, the removal of market support systems will probably lead to a significant reduction in sheep stocking levels in the Western Isles and the Highlands. In the better areas, such as the east-coast arable farms, farming is likely to become more technological, although not necessarily more damaging to the environment.

According to Tom Brady, the former chief executive of the Scottish National Farmers' Union, farmers are well aware that change is in the offing, and are ready to accommodate them. 'A lot of farmers in Scotland realise that this is what society wants. We are in an entirely different atmosphere. If that is the way they want to go, why should we resist them?'

By and large, farmers have had to be remarkably resilient over the years. When they were asked to produce more, they did, and now that they are being asked to produce less, and to restore some of the 'interest' to the natural environment around them, they are doing so. At the same time, many have embraced the call to diversify, and chalets and golf courses have sprung up all over the country.

The Way Ahead

The Scottish writer Jim Hunter once summed up the agriculture industry by saying it disposed of huge quantities of subsidies, made food bland, tasteless and unappealing, damaged the natural environment, and brought mad cow disease, *E. coli* and the like.

It was a simple and damning indictment, but not inaccurate. In an ideal world, the quality of the food we eat should be guaranteed. We should not have to pay a premium, as we do at present, to enjoy good, wholesome meat, or fruit and vegetables produced without chemicals. But we will be asked to do so for the foreseeable future.

The removal of tonnage- and headage-related subsidies could play an important role in bringing about environmental improvements. And there are sound reasons to suggest that subsidies should be paid on quality and not quantity, and that payments should be higher than at present and skewed towards environmental benefits. Farmers deserve to be paid more to be stewards of the natural environment.

For some farmers, anything that is not food production is not real farming, but that view must change. Economic reform should not lead to a major withdrawal of agriculture from the uplands, but to a new vision for their management. Something like the existing ESA schemes could be expanded to support traditional farming methods, extensification and long-term set-aside to allow wildlife habitats to be recreated.

We should also be open-minded about new uses for marginal land. It is possible that research already underway may lead to the farming of cashmere goats and wood being grown for use as a sustainable source of power generation. We should not see the current land uses in the uplands as the only options. In fact, both these examples would be viable now if all the subsidy available was not devoted to sheep and cattle.

The public's thirst for change has already been confirmed by a survey carried out by economists in Aberdeen who found that people were willing to pay an average of £55 per household per year to achieve an improved landscape. Those who took part in the study were shown drawings of three Southern upland landscapes, ranging from the current landscape, which is well grazed by sheep and lacks heather, trees and wildlife, to a final landscape in which sheep numbers have been reduced, and woodland and wildlife have been enhanced. Most people chose the latter option. In another study, householders said they would prepared to pay an average of £11 extra per year in taxation to support environmentally sensitive areas. This would give conservation benefits of £22m, compared to the current expenditure on the scheme of around £1m, or £1.20 per household.

Finally, another very fundamental change must take place. At present, integrated land use management does not exist in Scotland. Forestry and agriculture have separate assistance schemes, price support mechanisms and the agri-environment subsidies are largely for agriculture. The Forestry Authority pays grants for native species, community forests, commercial species and, in just one case, for farm woodlands. There is no tradition of farmers planting and harvesting trees – something which is commonplace in Scandinavia – and the EU does not allow most set-aside land to be used for trees. Equally, there are no incentives to encourage sound management of the soil at a time when soils are becoming more acidic, organic material is declining and the soil microflora and fauna are being reduced.

Better integration could be encouraged through the right kind of subsidies, and through community ownership of land. It is also likely that the removal of some agricultural subsidies would lower land values and open up more land to tree planting. This would end the situation in which most commercial timber in Scotland is grown on soils classed as unsuitable by the Forestry Commission.

8

·

Energy

·

You need more heating as you head up the kingdom away from
Essex. Luckily, we have thick walls and our own wood.
  Sir Nicholas Fairbairn, speaking of his home, Fordell Castle

A Question of Home Comforts
·

ANNIE MCINTOSH USED TO turn up the gas oven each
morning and leave the kitchen door open in an attempt
to heat the bathroom across the corridor. Her council flat
in the east end of Glasgow was so cold that she would leave
on the gas fire in the living room, two electric fires in the
bedroom, and put on layers of clothes. The windows in her
home on the edge of London Road were so ill-fitting that
she used towels to stop the rain coming in, and to keep out

the cold draughts. The tenement block was centrally chilled by the common stair.

Annie, 71, recalls her old, cold life with a cheerful laugh and a shake of the head. She was born in the area, and returned there 38 years ago to bring up her own children in a typical red sandstone tenement. As a young mother, she managed with two coal fires which were kept going all day in winter. Later in life, like tens of thousands of others in Glasgow, she became a victim of fuel poverty. Her family helped pay the bills, but the gas and electric heaters which had replaced the coal fires were unable to cope with the Glasgow winter. Her quality of life did not improve until 1997, when the Lilybank council properties were comprehensively refurbished.

The old roofs were replaced, new insulation was installed, and a two-inch thick polymer jacket was pinned to the outside walls. Only a single layer of the original sandstone, like a glimpse of petticoat, can be seen below the new exterior.

Annie's leaking windows have been replaced by double-glazed units, and her mix of inefficient fires has been replaced by thermostatically-controlled radiators in every room. There are new doors on the close which seal the stairway, and an entryphone improves security. The hot water and space heating for all 186 flats in the block is supplied by one gas-fired unit housed in the back court, and her monthly heating bill is now £14, regardless of the level of use. Before the transformation, her gas bill for March, the coldest period of the winter, was around £120.

'It's so different,' said Annie. 'I enjoy the heat and the new windows. Before the refurbishment the windows were gaping and it was like a storm blowing in. One of the walls in the bedroom was damp. I used to put on a foot warmer which stopped my legs getting burnt from sitting so close to the gas fire.'

Similar projects are being carried out around Scotland, and yet 750 000 households still suffer from fuel poverty. Transforming these households would achieve obvious social and environmental gains: Annie's quality of life has improved immeasurably; the refurbishment provided local employment; and – coincidentally as far as the tenants are concerned – Scotland's emissions of CO_2 were reduced because the flats are burning much less power. Progress, however, is limited by a lack of funding.

The work on Annie's house was a local authority initiative which was carried out by a private contractor, and by the Wise Group, a charity which has been providing employment and improving rundown areas of the city since 1984. It works in partnership with the council, government agencies, the EU and local communities, and in thirteen years has insulated over 115 000 homes, resulting in energy savings of £5m. Other parts of the company have planted 800 000 trees, and improved the environment outside 30 000 homes.

Tom McKenna, the supervisor responsible for the Lily-bank scheme, once worked on the Channel Tunnel but says he gets much greater satisfaction from helping people into employment. 'I think we are giving people a second chance,' he said. 'The people who work for us are not being forced on to a government scheme, they are here because they want to be here, and that makes a difference. The best news I can have is when one of the guys comes in on a Friday and tells me he is leaving because he has a job.'

Tom's comments, and the improvement in Annie's life, neatly illustrate the overlooked truth that the environment is about everyday life in the cities, every bit as much as wildlife and wilderness.

An Overview

We all consume energy, and we all play a part in local, national and global pollution. The coal, oil and gas which we require to cook, heat our homes and fuel our cars release pollutants which cause asthma, acid rain and global warming. The more cars there are, the more appliances we have in our homes, the more of us there are, the worse the situation becomes. The way energy is used is one of the key environmental problems in this country and many other countries, and the potential for change is great. Scotland is not using energy in a sustainable way, but it could be!

We need to replace polluting sources of energy with clean

ones. We need to swap coal-fired power plants for energy efficiency and wind power. We need to reduce the level of air pollution from traffic, through controls on car use and improvements in public transport. We need to invest in energy efficiency and end fuel poverty. And, many people would argue, we need to stop producing large quantities of radioactive waste.

The technologies necessary to achieve these changes already exist. They do not require development. For example, we lag behind Scandinavia, the USA and Canada in housing standards. New buildings in the north-east of the USA, which has a more extreme climate, use one third of the energy required by the average new house in Scotland. One Chicago builder, in a city famed for its fierce winters, guarantees that the annual heating bill for new, detached houses will be less than £120 a year. The same figure would not begin to provide adequate heating in a damp, poorly insulated council flat in Glasgow for more than a few months.

Where the housing stock has not been upgraded, people are trapped by fuel poverty, with unsustainable demands on their spending power and the consequent effects of ill health, which costs the National Health Service large sums of money in the treatment of respiratory diseases, heart and cerebro-vascular complaints. Families on low incomes could be provided with energy credits, and local authorities could improve the housing stock. It has been estimated that it would cost between £200m and £250m a year to tackle the worst 800 000 dwellings over a ten-year period, but Government funding for energy efficiency measures is far below that level.

According to the charity Energy Action Scotland (EAS), a total investment of £2.5 billion, in heating systems and energy efficiency measures, would create 50 000 jobs, and produce savings of almost £500 000 a year. The payback period would be seven years. Similarly, the 1993 Scottish Energy Study found potential energy savings of 50 per cent in the domestic sector, 10 per cent in industry and 40 per cent in transport. Transport alone consumes almost one third of the delivered energy in Scotland, and accounts for 30 per cent of CO_2 emissions.

To some extent, our continuing profligacy can be explained by the fact that Scotland has for long been rich in the fuel used to produce energy, with large reserves of coal, oil and gas, and a well-developed nuclear industry. We are also rich in alternative sources of energy, including wind and waves. But for the forseeable future, power generation will continue to be dominated by fossil fuels and nuclear power, which together supply over 90 per cent of our needs. The power derived from renewable sources accounts for less than 10 per cent of production, and 6 per cent of that is from the long-established hydro schemes.

A sustainable energy policy would lead to increased investment in renewables, and might also confront the proliferation of open cast coal mining, which is a blight on the landscape, a cause of pollution and a limited source of employment.

More radically still, some environmentalists would like to see personal taxes being replaced by taxes on resources. For example, FoE Scotland has suggested that a Scottish parliament could partly replace local business taxes with an industrial energy tax, which would act as a major incentive to invest in energy efficiency. Other schemes might lead to the individual no longer being taxed on the size of his home. Instead, he might pay more for fuel and consumer products. The development of windfarms in Scotland is already supported by a levy of 0.25 per cent on electricity bills, and £1 is taken from every domestic customer to pay for energy efficiency measures.

In the longer term, radical changes in the way we produce power will become inevitable as oil and gas run out in future decades, and the world coal reserves in centuries.

Unsustainable Lifestyles, and Environmental Space
International studies suggest the lifestyle of the average Scot is responsible for around eight tonnes of CO_2 per year, with national emissions of 43 million tonnes. One analysis of the energy crisis suggested we should reduce that figure to

5.4 tonnes by 2010, and to 1.1 tonnes by 2050. The latter figures are calculated using the concept of 'environmental space', which was developed in the Netherlands as a measure of sustainability.

The space referred to is the amount of any resource we can expend, or pollution we can emit, without diminishing the opportunities of generations to come. In the field of energy, the greatest concern is the emission of CO_2 from burning fuels, so CO_2 is used to determine our environmental space. If the figures proposed for Scotland were to be met, it would mean reducing carbon dioxide emissions by 38 per cent by the year 2010, and 87 per cent by the middle of the next century. The figure would be even higher if it took into account the flaring of surplus gas on North Sea platforms.

The good news is that the targets are smaller in Scotland than the rest of the UK, due to the size of our nuclear and hydro resources, and that our personal emissions have dropped from 9.64 tonnes in 1989, largely due to the restructuring of our industrial base and the switch from coal to gas-fired power generation. The single act of closing the Ravenscraig steel mill in Motherwell, in 1989, reduced the Scottish demand for industrial coal by 75 per cent and Scottish CO_2 emissions by 6 per cent.

Scotland should also take heart from the fact that it is well ahead of the world's most polluting nations. While the average European produces 7.3 tonnes of CO_2, each North American accounts for a massive 20 tonnes. Compare that with the output of the average African, who is responsible for 1.03 tonnes. The figures make the point that it is the developed world which is responsible for climate change.

Energy in the Home

Around one third of Scotland's 2.1 million households are unable to heat their homes to an adequate level, with the poorest people often using the most expensive forms of electric heating. At the same time, relatively low fuel prices mean there is no great incentive for the more affluent to invest in energy efficiency.

Household demands consume 25 per cent of our energy production, and produce 22 per cent of our CO_2 emissions. Several studies have suggested that a 50 per cent reduction in energy use in the home is attainable, and one expert was moved by such statistics to call Scotland 'energy illiterate'.

Better building design, for example, might result in 90 per cent of domestic heating needs being met from passive solar energy. Our building regulations today are no better than those that existed in Scandinavia in the 1930s, and people tend to over-estimate the effect of visible measures, like double glazing, and under-estimate the effect of invisible measures, like cavity wall insulation.

But there have been few serious attempts to address the situation. In 1996, the Home Energy Conservation Act required councils to assess the energy efficiency of the housing stock, and to draw up plans to improve the situation. The aim was to provide warmer homes at lower cost and to reduce energy demand. The Act advised an improvement in energy efficiency of 30 per cent over ten years, but local authorities maintain any action is limited by a lack of resources. Meanwhile, 400 000 Scottish homes continue to have a National Home Energy Rating (NHER) of 2 or less, out of a possible rating of 10.

When I had my former home, a 40-year-old bungalow, tested in 1992, it was given a miserable rating of just over 3. I was told it was releasing around 23 tonnes of CO_2 a year into the atmosphere of central Scotland, partly due to an inefficient oil-fired boiler and a lack of insulation. I was also told that spending £1000 would drastically improve its efficiency, and that the investment would be repaid in reduced heating bills in three years.

The Utilities

Over-capacity in Scotland has delayed the development of renewable energy, although both ScottishPower and Hydro-Electric are showing an increasing commitment to improved environmental standards. Yet they remain some way short of the new attitude of the power generating industry in the USA, where it is accepted that a unit of energy saved can be more profitable than a unit generated. Some American companies have chosen to invest in energy efficiency measures, such as giving away low-energy lightbulbs, when faced with the alternative of building more capacity to cope with rising demands. The largest power utility in the USA closed its power station construction division in 1993, and took a major step towards a mix of renewables and energy efficiency. But in the UK, selling electricity is what the utilities aim to do, and big users are still rewarded by lower rates.

The situation is not helped by the fact that the total generating capacity in Scotland is double the maximum demand, a situation which resulted from inaccurate projections of the growth in demand calculated in the 1970s. Under existing agreements, both utilities are contracted to take all the power produced by the Hunterston B and Torness nuclear power stations until 2005, regardless of the fact that cheaper sources exist elsewhere.

Around 80 per cent of Scottish demand is met by just five stations: the two nuclear plants; the Peterhead dual oil and gas plant; and the coal-fired stations at Longannet and Cockenzie. ScottishPower, meanwhile, is aiming to step up its exports, and plans to increase its use of coal from 2.2 million tonnes in 1994 to 5 million tonnes a year by 2000, which will lead to higher emissions of CO_2.

At the same time, the utilities show no great enthusiasm for so-called demand side management (DSM), which potentially offers significant energy savings. One exception occurred in Shetland, where power is supplied by an ageing oil-fired station with a poor pollution record. Faced with the problem of demand outstripping supply, Hydro-Electric responded by installing switches which alternated the time when storage heaters came on, and thereby spread the peak-period demand. It also offered free energy audits, and gave grants towards insulation and draught-proofing.

Hydro-Electric has shown limited interest in renewables – other than its long-established hydro resource – but is developing combined heat and power systems, which have been common in Scandinavia for decades. These schemes are much more efficient than conventional coal, oil and gas-fired stations because they use the steam which turns the turbines to supply hot water and space heating to groups of houses and businesses. The existing UK target for CHP schemes is 5000MW by the year 2000, which would reduce CO_2 emission by 5 million tonnes.

Recent environmental advances by ScottishPower include the reduction of NO_x emissions from Longannet and the marketing of pulverised fuel ash, which was formerly dumped, as a construction material. It is also committed to wind power, and has schemes throughout the UK. ScottishPower's polluting outputs are below the UK average, but its emissions of SO_2 and CO_2 have increased slightly in recent years due to its increased export sales to England and Wales. Hydro-Electric's CO_2 outputs are also higher now than they were in 1992. In the longer term, the behaviour of both companies may have to change drastically if they are required to account for the 'externalities' of power generation, such as the effect on global warming of CO_2 emissions from power plants.

A directive is being studied in Europe which could result in these costs being internalised for industry. It has been suggested that 1p worth of environmental damage is caused for every kilo of CO_2 emitted. When the sum is applied to Hydro-Electric, it reveals that the company is responsible for £150m worth of environmental damage each year, amounting to a significant 15 per cent of its turnover.

Privatisation

Fuel prices have dropped since electricity privatisation, and deregulation has encouraged competition and allowed customers to shop around for the best deal. Business has therefore reduced its costs, but, as a result, its energy use per unit of production is higher than it could be, and companies see less need to invest in energy efficiency measures. Put another way, the fall in fuel prices has resulted in a 5 per cent increase in the amount of CO_2 being emitted by industry.

For every 10 per cent fall in energy prices, energy demand is estimated to rise by 4 per cent. The reduction in gas prices since the privatisation of British Gas is said to have resulted in a 22 per cent increase in the use of that energy source. The conclusion reached by the Association for the Conservation of Energy (ACE) is that privatisation has been disastrous for energy conservation and the environment. The situation may have to be addressed by incentives, or regulations, to drive business investment in conservation.

Further liberalisation of the energy markets from September 1998 could reduce the incentive for domestic consumers to invest in energy efficiency. From then, the individual consumer will be able to choose the source of his power, and is likely to go for the cheapest tariff, rather than attempting to reduce consumption. However, there have been indications that some households may be prepared to pay a small premium for 'green' electricity. FoE Scotland has suggested that electricity bills should be marked like food labels and contain information stating how much CO_2 and SO_2 the company's power plants produce.

Environmental considerations can only become more pressing for the power giants. The Scottish utilities have already started to receive questionnaires from ethical investment advisers, asking for very detailed information on, for example, CO_2 emissions. The companies obviously stand to gain if they are clean enough to be listed as environmentally sound investments.

The Resource

Non-Renewable Energy

The current regime of over-production is certain to last well into the next century. Hunterston B nuclear power station (producing one quarter of Scotland's electricity) is due to survive until perhaps 2010, and Torness until 2023. Their loss could be compensated for by a combination of reduced exports, increased energy efficiency leading to a reduction in demand, and new renewable capacity.

Coal

Coal production peaked at 43 million tonnes in 1913, and has been declining ever since. In 1997, just under seven million tonnes were being produced, with four million tonnes coming from open cast operations. In 1996, there were 61 open cast mines, with 50 applications in the pipeline. The industry relies heavily on contracts for the two ScottishPower stations at Longannet and Cockenzie.

Over the years, coal consumption has been dominated by the electricity industry, although domestic consumption has also declined. Cheaper fossil fuels have caused the decline, although substantial reserves remain below ground. Coal production, however, is a polluting business at the production and consumption stages, and mine abandonment has become one of the principal causes of river pollution.

Oil and Gas

Oil shale was developed by James 'Paraffin' Young in the 1860s, and paraffin was a major source of energy for a century. Red shale waste threw up huge man-made bings which still dominate parts of the central lowlands.

Offshore oil and gas were found in Scottish waters in 1969, and their rapid development led to the first landing of oil in 1975, and a production peak in 1985. North Sea oil is low in sulphur and ideal for transport fuels. Production is now moving towards the fields west of Shetland and the so-called Atlantic Frontier area around Rockall, with Scottish oil likely to flow for at least another thirty years. Technological developments in the next few decades could extend that period considerably.

The oil business is a polluter from the first stage of exploration, through the burning of the fuel, to the final disposal of decommissioned platforms. However, the BP Sullom Voe oil terminal on Shetland deserves mention as a major industrial installation which serves as a model in terms of environmental monitoring.

While oil companies have been implicated in attempts to disrupt international agreements on global warming, the more enlightened players are looking to diversification.

Statoil, the Norwegian state company, claims it has begun the process of changing from an oil and gas company into a firm with a broader role encompassing renewable energy.

The role of the oil industry in pollution is significant: the 130 million tonnes of oil extracted from the North Sea in 1995 released 319 million tonnes of CO_2 when burned, which was seven times the total Scottish onshore emissions. Gas production in the same year reached an all-time high, accounting for 159 million tonnes of CO_2, while the gas flared in the North Sea emitted 4.6 million tonnes.

Nuclear Power

In 1990, around 400 nuclear power plants, including eighteen in the UK, produced 16 per cent of the world's electricity. Those who favour the industry believe it should be expanded to help reduce the polluting effects of conventional power generation. But nuclear power has a poor public image, and its generation of highly radioactive waste entails environmental risks for millennia to come. High-level waste can remain dangerous for 250 000 years. If the material is reprocessed, it generates large quantities of intermediate and low-level waste, and much of the high-level material remains. The Royal Commission on Environmental Pollution stated in 1976: 'It would be irresponsible and morally wrong to commit future generations to the consequences of fission power on a massive scale unless it has been demonstrated beyond reasonable doubt that at least one method exists for the safe isolation of these wastes for the indefinite future.' Seems fair.

At present, the industry is responsible for around 40 per cent of Scotland's electricity, and up to 50 per cent during the summer months. Most of that output comes from Scottish Nuclear's advanced gas-cooled reactors at Hunterston B on the Clyde coast, and Torness on the Forth. The third nuclear plant, at Chapelcross in Dumfries and Galloway, is run by British Nuclear Fuels for military purposes and supplies power to the English and Welsh grid as a by-product. The major environmental impacts are caused by the discharge of cooling water, the disposal of nuclear waste and emissions of radioactive material into water and air.

However, levels of radioactive pollution from everyday operations in Scotland are small, and discharges from nuclear power stations account for only 0.1 per cent of the average annual dose of radiation which we experience, compared with 51 per cent from naturally occurring radon and thoron gases.

Nonetheless, there are concerns about the possibility of an increased incidence of childhood leukaemia around some nuclear sites. SEPA has recently expressed concern about emissions from the Dounreay reprocessing complex, and about the cocktail of radioactive material dumped in a waste shaft which was opened at the site in the 1950s. It contains low-, intermediate- and high-level waste, and it was revealed after an explosion in 1977 that the contamination was six times greater than previously admitted.

At present, the low-level waste produced in power plants, including items such as gloves and laboratory equipment, is disposed of in a thirty-acre landfill site at Drigg in Cumbria, which also deals with radioactive waste from industry and hospitals. The last government gave Nirex the difficult task of developing a deep underground repository for inter-mediate-level waste (which is currently stockpiled), which would have to hold the material for many thousands of years. Its choice of Sellafield in Cumbria, near the large reprocessing plant, has proved ill-advised because of the geology, and scientists may look again at the possibility of a site in the hard rock near Dounreay. Environmentalists, meanwhile, recommend above-ground, on-site storage until such time as a safer method of disposal can be developed.

Not surprisingly, nuclear power, and in particular the issue of storing or reprocessing the wastes, is regarded as a hugely important issue by many environmentalists. It is a more expensive source of power than coal, oil or gas, but, like renewable energy, is offered market protection. Before privatisation, billions of pounds of debt were written off.

Today, the regulations governing the industry are strict, and a far cry from the 1950s, when the planning application for Dounreay was hand-delivered to the local county council in the morning, and collected, duly signed, in the afternoon. According to polls, more than 80 per cent of Scots were opposed to nuclear reprocessing at Dounreay, while more than 60 per cent of the people in Caithness objected to the plant bidding to reprocess spent nuclear fuel from around the world. In the summer of 1998, the Government decreed that no new contracts were to be signed, and reprocessing was to stop by 2006.

Renewable Energy

The development of a renewables resource in Scotland is an important part of the complicated jigsaw of a sustainable energy policy. Green power is supported by the Government's Scottish Renewables Order (SRO), which obliges ScottishPower and Hydro-Electric to source a small percentage of their output from renewable schemes.

The first Scottish projects, announced in 1994, included wind, chicken litter, and refuse powered schemes. In 1997, another sixteen projects were approved – seven wind farms, nine small-scale hydropower schemes and one biomass scheme. However, the existing renewables resource is dominated by the long-established hydro schemes, which produce 1200MW. Wind power accounts for just 43MW.

In Scotland, the SRO aims to provide 150MW (one tenth of the UK target) of new renewable capacity by the year 2000, or about 5 per cent of electricity demand. However, there are numerous problems with the current system, including the fact that the SRO is limited to wind, hydro, waste-to-energy and biomass schemes, and does not support wave, tidal or solar power. It is also based on the price per unit of power produced, and therefore encourages the cheapest schemes, which are sometimes poorly planned, or, in the case of wind farms, situated in the most sensitive locations to take advantage of the highest wind speeds. Wind farms on hill tops and ridges often fall victim to the Nimby (not in my back yard) syndrome, and can be defeated by the planning process, even after being awarded a contract under the SRO.

The current Scottish renewables resource has been

estimated at 29 000MW. According to FoE Scotland, 50 per cent of the country's electricity could be produced from renewables by 2025. With the necessary will, changes can be made on a large scale. Japan has a programme to fit 70 000 photovoltaics (solar energy) systems to roofs, with the aim of establishing a viable industry with a massive export potential.

Peat

Peat has traditionally been used domestically, and has only been exploited on a larger scale relatively recently. Current levels of exploitation are thought to be less than the natural rate of production, although peat from Caithness is sent to Sweden as a power-station fuel. The main use of lowland peat is for gardening products.

Wind Power

Modern wind power is a development of the windmills of yesteryear which were used to grind wheat and pump water. Before the industrial revolution, there were around 10 000 windmills and 20 000 water mills in Britain.

Small wind turbines are now widespread, and Scotland has three significant windfarms, one run by ScottishPower at Hagshaw Hill, Lanarkshire, which produces 15MW (enough electricity for 13 000 homes), and a more recent one at Windy Standard, Dumfries and Galloway, with an output of 21MW. A third site, at Novar in Sutherland, has recently become operational. Wind power has developed faster than any other new renewable in Scotland, due to the massive 'wind regime' enjoyed all over the country.

Globally, wind power matched nuclear power in 1995 in terms of new capacity, and it is much cheaper than splitting the atom. Four million Europeans get their power from wind, but very few of them live in Scotland, which has 25 per cent of the European wind power potential. The current capacity in Scotland – ranging from the three big wind farms to the micro-machine which provides lighting for Glen Affric Youth Hostel – is 42MW, compared to 270MW in England and Wales. There is great potential for further development, but the current SRO regime aims for just 150MW of renewable energy by the year 2000, while the Scottish Office estimated that no less than 8000MW could be supplied by the resource.

Windfarms offer a rental to landowners, have a very limited effect on wildlife and are not necessarily an eyesore if they are sited properly. Noise is one of the key concerns, although the technology is constantly improving. The average turbine is now said to produce 35–45 decibels of noise, compared to the 35 decibels of a 'quiet bedroom'. Today's wind turbines are said to be three times more efficient than those erected in Denmark in the 1980s. The Danes have led the technology for some time, and one recent development outside Copenhagen has turned power supply into a community business. Around 800 shareholders have invested in seven efficient turbines to produce the power for 3000 families.

Vestas, the leading wind turbine company, employs 1200 people and exports its equipment and expertise abroad, while Scotland fails to develop or fund environmental technologies, but spends vast sums to attract Japanese and Korean computer companies.

The next step for wind farming in Denmark is the development of wind farms at sea. Plans are in hand to have 4000MW of capacity installed 5–10km offshore by 2005.

In the UK, where the industry is far more advanced south of the Border, there is a proposal to site twenty-five wind turbines off the coast of Great Yarmouth. ScottishPower, meanwhile, has opened a new 25-turbine windfarm in Donegal, and operates sites in Cornwall, Lancashire, Northern Ireland and Wales.

Hydro Power

Hydro-electricity is the long-established source of water-derived energy, although tidal and wave energy are likely to be key sources of power in the future. Hydro power has been well developed in Scotland for several decades. No new large-scale enterprises are likely, although the SRO supports the development of small 'run-of-river' schemes, mainly on private land and producing small amounts of power, with any excess going to the National Grid. Hydro power is the cheapest form of electricity in Scotland.

Wave Power

Many schemes are still at the experimental stage, but wave power has enormous potential, and, like wind power, offers more power in winter when the demand is highest. In theory, the UK potential is equal to the current total electricity demand, but increased government funding is needed to give the technology a much-needed boost.

There are no tidal systems proposed in Scotland, but one onshore wave power device has been installed at Islay, and offshore systems, notably the famous 'nodding duck', invented in Edinburgh by Professor Stephen Salter, are under development.

Wave power got off to a false start in the UK when the programme was abandoned in 1982 amid allegations of dirty tricks by the nuclear lobby. This may have had something to do with the fact that the technical figures for Salter's prototype duck were worked out at the Energy Technology Support Unit, which is based at the headquarters of the UK Atomic Energy Authority in Harwell. Professor Salter, who began work on the duck in 1974, eventually overcame the setback and remains at the cutting edge of wave power, but aims to be 'the last at sea'. He is working towards the creation of one huge 500-tonne solo duck, to be stationed off the Outer Hebrides. It would produce power at 5p per kilowatt hour, which is less than nuclear power and other wave devices. It would have an output of 3MW, enough for a community of a few thousand people, and would be 24m wide and 10m high. He aims to create an energy device so technically sophisticated that it will not require on-shore monitoring, although engineers will be able to work inside it.

Biomass

The word biomass covers a large range of different fuel sources, including short rotation forestry (normally willow or poplar); wood waste from forestry, sawmills and construction; sugar crops, including sugar beet and Jerusalem artichoke; starch crops such as wheat and maize; oil crops such as rapeseed or sunflower; agricultural wastes like straw and slurry; and municipal and industrial waste, including residues from the food and drinks industries. There appears to be considerable economic potential linked to a short, 3–5 year, rotation of trees on good agricultural land, or a 12–20 year rotation on upland sites.

Biomass remains the major fuel for three quarters of the world's population, and was only replaced by coal in industrialised countries in the last 200 years. The energy from biomass is normally released by burning, but can also involve the production of methane gas from decomposing waste in landfill sites. The advantage of burning biomass instead of fossil fuels is that the CO_2 released is equivalent to the amount that was taken up by the biomass as it grew, so there is no nett release.

Although such fuels have only recently been supported by the SRO, they are widely used in Norway and Sweden in district heating systems. At one site outside Stockholm, coal and oil have been replaced by powdered waste and straw which heat around 25 000 homes.

Under the SRO, the burning of refuse and industrial waste, including plastics derived from oil, is allowed, and Dundee has been given approval for an energy scheme based at an incinerator site. However, many environmentalists do not consider the burning of waste, which could be recycled, to be a genuine source of renewable energy.

Solar Energy

There are three types of solar energy: passive solar, which involves building design; photovoltaics, which involves the transformation of the sun's energy into electric current; and active solar heating, which uses solar collectors to capture the sun's energy to heat air or water for use as space heating and/or a hot water supply.

Photovoltaics, which is used on solar-powered calculators, is an important option for Scotland. The panels can be used on offices, factories and homes. The potential, according to one study, is the equivalent of one quarter of the present electricity demand. The panels are, of course, silent, have no emissions, and the planning consideration is confined to their visual impact.

Once again, Scandinavia is ahead of the UK. In Sweden, there is even a field full of solar collectors which heat an underground cavern filled with water and distribute it through district heating systems to heat a small town over winter. In the USA, the government is playing its part through market regulation, and by providing annual research funding of £300m, compared to £18m in the UK.

9
·

Transport

·

Motor cars were to be the great means of propulsion in the coming age ... And the unending roads are now made, with their tarmac shiny surfaces, their curb-bound borders, their loathsome motor pumps, their destruction of every lovely turn and bend, and flowers wither wherever they are driven.

Lady Frances Balfour, churchwoman, suffragist and writer, *Ne Obliviscaris: Dinna Forget*, 1930

THE MOTOR CAR IS both heaven and hell; it is idolised and vilified, glamourised and despised. It is a status symbol, a sex symbol and an extension of our personalities. Cars and their owners are clichés. Poor men don't drive Porsches.

Cars are the ultimate form of transport. They can be bought and sold and take us anywhere – just us and our loved ones. We do not have to interact with other people, except to avoid bumping into them. We can choose where to go, and when to go. Cars confer power and freedom, and they make us drool and wring our hands. Nearly all of us like them, most of us own them or want to own them, and, to keep the whole merry-go-round going, we produce 48 million of them every year, or about one every second. In eight hours, there are another 40 000 cars, in a day another 100 000. There are a great deal more of them than Queen Victoria could have imagined, when she observed: 'I am told that they smell exceeding nastily and are very shaky and disagreeable conveyances altogether.'

Throughout the world there are more than 500 million lorries and cars, one for every ten people, and that figure is set to double to one billion in the next twenty years. According to the CBI, congestion is costing British industry £15 billion, and it is impossible to build new roads fast enough to cope with the forecast traffic growth. Should we worry in still green, and quite pleasant, Scotland?

The answer is an unequivocal Yes, and there are numerous compelling reasons for saying so. Even in Scotland, the relationship between us and our cars is showing signs of strain.

Ironically, when the internal combustion engine was invented, it was thought that it would ease congestion and reduce pollution in Victorian Britain, where three million horses each produced three or four tonnes of dung per year, and consumed the produce of 15 million acres of farmland. Kipling, for one, was convinced, and called the motor car 'a civilising agent, a moral force, one of the dynamic factors of the great march of progress'.

But, as early as 1907, Asquith – a Prime Minister apparently blessed with greater insight than Margaret Thatcher and John Major had nine decades later – predicted that the motor car would become 'a nuisance'. Another ninety years would pass, however, before any government would signal decisive action to curb the growth of the car culture.

There are numerous key dates on the journey to our present situation. In 1938, the County Surveyors Society concluded that what Britain needed was 'high-speed, multi-lane roads exclusively for use by motorised vehicles'. In 1955, we celebrated the first 100 years of motoring, the first motorway was opened in the 1960s, and the first 1000 miles of motorway were completed by 1971.

Today, we are re-examining the whole relationship. The internal combustion engine which was meant to be our servant, has become our master. Our cities have become places for the movement and parking of vehicles, not places to play, to converse, or watch the world go by. Physicians, as well as conservationists, are concerned. Traffic and roads do, of course, create congestion and destroy precious wildlife habitats, but it is possible to argue that for health reasons alone, we need to change the way each of us gets from A to B.

Only three diseases are on the increase in Scotland – AIDS, lung cancer and asthma. The latter, affecting up to four million people in the UK, has been linked to the air pollution from traffic which is said to cost £1 billion a year in treatments in UK hospitals.

Throughout Britain, more than 19 million people are exposed to air pollution which exceeds EU guidelines. Drivers and their passsengers in London breathe levels of carcinogenic benzene which are close to the limit for occupational exposure to the gas. It is estimated that around 2000 Scots die each year as a result of traffic pollution, around four times the number of people killed in road accidents. Transport is the single biggest cause of air pollution, the fastest growing cause of global climate change, and the problems are increasing with the seemingly inexorable growth in car ownership, regardless of cleaner technologies.

Cars and lorries, diesel and petrol, spew out a cocktail of potentially dangerous pollutants. We know exactly what some of them do to us, but are still finding out about the potentially lethal effects of others.

1. Carbon monoxide (CO) causes drowsiness and headaches and impairs blood-oxygen levels, particularly in the young and the elderly. In confined spaces it is lethal.
2. Nitrogen dioxide reduces lung functions, exacerbates

asthma and helps create acid rain which damages freshwater fish and plant life. It is also a component of ground-level ozone which irritates the lungs, aggravates bronchitis and heart disease, and damages crops.

3. Hydrocarbons cause coughing and drowsiness, and include the carcinogen benzene, which has been linked to leukaemia.

4. Sooty particles (PM10) from diesel have been linked with deaths from respiratory and heart ailments.

Technological advances mean that exhaust gases are becoming cleaner, but the sheer growth in vehicle numbers means there is no technical fix. In the 1980s, car manufacturers said everything would be fine when catalytic converters were fitted as standard, but the same technology has been on American cars for more than a decade, and does not stop the smog in Los Angeles.

If health concerns are not sufficient to prompt action, then congestion should be. When the public are asked to cite environmental problems, transport often comes out top. This helps to explain why the Labour Government is taking action. Planners and politicians respond to public opinion, and there may already exist a mandate to take radical action on Scottish transport.

Change is long overdue, and could reverse the trend of the past 100 years. It has long been claimed that building roads is good for the economy and good for freedom of choice. Successive Tory transport ministers in the 1980s and 1990s said people wanted more roads, wanted to be able to exercise their individual right to drive a car, wanted better access to the good things in life. In 1994, the British Road Federation was still insisting that better roads would boost the economy and create long-term jobs. But there is little evidence to suggest that building roads assists growth and efficiency. In 1977, a Government advisory committee on trunk roads concluded that improvements to the trunk road system had only a limited effect on industrial location and growth.

Between 1984 and 1994, the Department of Transport managed not to commission any studies on the economic benefits of road building, but in 1994, the trunk road advisory committee said, categorically, that new roads generated increased demand for road travel and increased the number of road journeys. This statement should have made nonsense of the last government's plan to turn the M25 into an eighteen-lane highway.

An even starker message emerged in 1994 from the Royal Commission on Environmental Pollution, which put forward 110 recommendations for change. It wanted a 40 per cent increase in fuel efficiency by 2005, and spending on the road programme to be halved, with the money being switched to a ten-year programme of public transport improvements.

Two years later, a damning report on Scottish road building plans was produced by the Scottish Wildlife Trust, which calculated that 92 current, or planned, road schemes would affect 254km of natural and semi-natural habitat, and damage 100 conservation sites. Another study suggested that road traffic in Britain killed between 30 and 70 million birds, 40 000 badgers and 5000 barn owls every year.

The death toll was not helped by the fact that throughout the 1980s, new road schemes were 'joining the dots' between environment sites, which had a relatively low development value because of their conservation status and were therefore cheaper to cross. EU legislation now requires an environmental assessment on all major trunk roads. Eventually, government action and popular discontent was translated into radical action on the streets, with direct action being taken at major road schemes in the south of England on a scale not seen since the anti-nuclear protests of the 1980s.

The extension of the M77, linking the Glasgow motorway system to Ayrshire and the south-west, became the first *cause célèbre* in Scotland for the colourful brigade of protestors which travels from one development site to another. The motivation, and the level of knowledge, of some of these environmental warriors may be questionable, but they are only the extreme edge of a popular revolt. They may want, for ill-defined reasons, to save a few unremarkable trees or a colony of natterjack toads, but like the majority of us, they also want better transport systems in our cities.

Current Trends
.

The trend in vehicle ownership continues upwards. In 1951, 14 per cent of all households owned a car, compared to 68 per cent today. In 1952, the average person travelled six miles a day, compared to 25 miles today. In the decade between 1960 and 1970, car ownership in Britain rose by a staggering 97 per cent, and has continued to grow at 30 per cent every ten years.

What of public transport? Between 1971 and 1991, the number of Scots travelling to work by car or motorcycle rose by 96 per cent, while the number using public transport, which is much less polluting, dropped by 48 per cent. The downside of these trends can be observed in London, where the average speed of traffic in the rush hour is 7mph.

What Do We Need in Scotland?
.

Scotland, it should be remembered, is a small country, and its transport system should reflect that fact. It is small enough for an integrated transport system planned at national level. And, despite all the concerns listed above, it is also several years away from the kind of problems experienced in London and other large English cities.

We are still at the stage of developing our trunk road network and we do not yet have unbroken dual carriageways or motorways between Perth and Inverness, Perth and Aberdeen, or Aberdeen and Inverness. We also have very small roads in the west Highlands coping with heavy tourist traffic in the summer months, and heavy fish lorries all year round.

In these circumstances, it would not make sense to halt all road projects, but it might make sense to say there should be no further motorway development in Glasgow, and it would make sense to suggest the development of sophisticated public transport systems in both Glasgow and Edinburgh, the only two seriously congested cities in Scotland.

The task has been rendered more difficult by the recent deregulation of buses and the privatisation of the railways. The deregulation, brought about by the 1985 Transport Act at a time of declining bus patronage, has had the effect of increasing the number of buses on the road, but decreasing the amount of bus usage. It has also made it difficult for authorities to exercise any control, while operators tend to plan in the short term in the interests of competition, rather than the interests of the passenger. The quality of the service has declined and new operators are unable to invest in vehicles because of high loan charges.

Rail privatisation was put in place in April 1994, when British Rail's functions were handed to 100 separate companies. Rail usage had been declining in Scotland, while an increasing amount of freight was being conveyed by road, with all the attendant drawbacks of juggernaut traffic. In 1994, 65 per cent of the total freight tonnage in Britain was shifted by lorries, compared to 6 per cent by rail. Sustainable transport planning requires something like a reversal of those figures.

Such drastic requirements have been recognised in other countries, with some European cities at least ten years ahead of Edinburgh and Glasgow. The evidence in Zurich, Stockholm and Freiburg, where it has been made more difficult to use the car, suggests that people will choose to use public transport if it is good. In Zurich, traffic lights change in response to on-board computers on buses, to give them priority ahead of cars. Pedestrians and cyclists also have priority over cars as part of a series of measures designed to make driving more difficult and irritating. The result? The car is left at home. The same city has begun reclaiming its centre and residential areas. Parking areas have been replaced with gardens, concrete has given way to flowers outside apartment blocks, the city centre has become a more people-friendly environment, and the transformation is visually striking.

One argument against such a transformation is that retailers will lose business, but the reverse may be true. Pedestrianisation can recapture the vitality of a city centre, and result in people relaxing and spending more time there. But such changes only succeed if there is a fast, efficient, clean and comfortable alternative. The system is so well integrated in Zurich that theatre, football and drama tickets include the price of transport on the mix of modern trams and buses serving the city. There is no need to buy a ticket on the bus. Compare these measures with the business of getting a bus in Edinburgh and Glasgow, where the information on bus routes and timings is primitive. Good public transport does not necessarily mean the diesel-belching, maroon double deckers of Edinburgh.

Transport solutions also have to consider rural areas, where many people have no alternative to the car. For instance, increases in fuel prices, designed to reflect the real cost of motoring and to encourage lower levels of car use, unfairly penalise rural car owners. Policies must be modified to fit rural areas, and to provide greater opportunities for access to goods and services in the countryside.

The prospect of movement towards some of the goals outlined above has improved markedly with the change of government, although the previous Conservative administration was making some effort to move away from the Car Society, and had cancelled 600 planned bypasses. The Labour Government has cut the roads programme further, and Malcolm Chisholm, the former Scottish transport minister, said the Government was committed to 'examining ways of restraining, or reversing, growth in car use'. Its goal is restraining car use, not ownership.

Labour has already, for example, vetoed the entirely unsustainable idea of a second Forth Road Bridge and the proposed Barnton bypass. The task facing the Government is to make you and me stop using our cars, an aim which is particularly difficult at a time when the cost of motoring has fallen by 25 per cent over the past ten years, despite the increased taxation on car ownership and fuel, which, in successive Budgets, has failed to curb car use.

A more practical tool in controlling the use of Scotland's 1.9 million cars may be the introduction of road-pricing. A charge of £1 for each vehicle entering Edinburgh could raise more than £60m a year, a large proportion of which might be used to fund public transport. Professor Phil Goodwin, the head of a government panel on transport, has suggested that voters will support town centre road tolls as long as the proceeds are spent on improved public transport. In September 1997, he said road pricing was the best means of raising large sums of money for public transport.

Edinburgh – Ahead of the Pack

The Scottish capital is one of the most beautiful cities in Europe, with an outstanding central thoroughfare in Princes Street. In a more progressive country, Princes Street would be pedestrianised; it would be a boulevard of street entertainment, gardens, flower sellers and outdoor cafes. It might have an elegant, light tram running through a leafy corridor at its heart, but it would have no cars and buses, no exhaust fumes, no road accidents. It would be, even more than it is today, one of the most photographed streets in Europe.

So far, the local authority has removed cars, but not buses, from one side of Princes Street, and even that modest achievement resulted in a 5 per cent reduction in traffic in the city centre. But if Princes Street is not yet a haven of peace, Edinburgh is, at least, leading the way in Scotland, and possibly in the UK, in its approach to better transport systems. David Begg, the transport convenor of Edinburgh City Council, wants to 'uncouple the link between economic growth and rising car ownership', and to prove that a successful economy and a good environment are not mutually exclusive. He also wants decision-makers to ask, not how to control traffic, but 'What sort of city do we want to live in?'

He has his work cut out. Car ownership in the city is currently growing at more than twice the national average and, while Edinburgh may aspire to be like Zurich, the local authority does not have the necessary clout to develop its own integrated transport system. It aims to reduce car use by 30 per cent by 2010, but will be lucky to do so. Other targets include increasing the share of public transport from 34 per cent to 41 per cent, and cycling from 2 per cent to 5 per cent.

Meanwhile, the public are well aware of the problems of congestion, but are poorly informed on the subject of air pollution and health. The levels of nitrogen dioxide found in the air in several busy streets, and also in Waverley Station, breach national limits on a regular basis. But in European terms, Edinburgh is likely to be in the cleaner half of the air pollution league tables which will be compiled under a new EU directive. Begg believes such a table will help win support for his radical ideas, and suggests that clean air may attract inward investors in the more environmentally-conscious future.

Edinburgh's approach to its transport problems is centred around integrated transport, planning and land use systems, and involves the expansion of bus 'greenways', which have taken away existing road space from cars. A metro has been ruled to be too expensive in the immediate future, but a guided bus network, known as CERT (City of Edinburgh Rapid Transit), may play a major role in the switch to public transport. There is also a need to create a 'bus of the future'. Such developments mean a return to serious planning decisions, and a move away from Mrs Thatcher's free market, when business and retail parks mushroomed on green-field sites.

Surprisingly, the city starts at a slight advantage because it has the lowest car dependency in the UK, with the exception of London. Around 48 per cent of people use a car to get to work in Edinburgh, compared to 50 per cent in Glasgow and 59 per cent in Manchester. It also has an unusually high level of bus use, with over 30 per cent of employees using buses to get to work, and an above average number of pedestrians, over 11 per cent, walking to work. Yet it has a minimal heavy rail network, and no light rapid transit system.

It benefits from the fact that it is European in style, with a high urban density and a large number of people living in the city centre close to the facilities they need. But it has not been a paragon of fine transport planning in recent years. The biggest commercial development in the city during the 1990s was the Gyle Shopping Centre, a greenfield development which created a typical, US-style out-of-town mall and business park. It was located next to the city bypass, and all the units on the site have large car parks. As a result, around 94 per cent of people using the centre travel by car. One study showed that the complex created 30 tonnes of CO_2 a day, and caused 5 per cent of the air pollution in the city. The Gyle is unsustainability exemplified. It is the antithesis of what forward-thinking planning authorities should be doing, but it was in keeping with the thinking of the day.

There must be no more Gyles in Edinburgh. The city may do better with the proposed Ocean Terminal at Leith, which is to be designed around public transport usage. The authority hopes that by the year 2010, more than 40 per cent of trips to the centre will be by public transport. The developer will suffer penalties if the goals are not reached. The city authority is also aware that residential development will have to be achieved in a more sustainable fashion, without producing more low-density commuter suburbs. It is considering a development to the south-east of the city which will have a more self-contained, urban character, with shops, employment, community facilities, and good public transport provision. Other plans include those detailed in the following subsections:

City Car Clubs

Edinburgh plans to launch City Car Clubs modelled on schemes found in Germany, Austria and Switzerland. These schemes operate on a form of shared car ownership in small communities. There is one car for every ten to fifteen households in the Marchmont area, with the vehicles available at any time for an hourly, daily or weekly fee. The schemes provide access, not ownership, and encourage the use of public transport. They also reduce residential parking pressure in inner urban areas. The evidence from Europe suggests that members become more aware of the relative costs of different forms of transport, and the individual mileage travelled by car drops by as much as 50 per cent. There are 3000 car club members in Berlin.

Car-free Residential Areas
In these housing areas, residents agree not to own cars. The result is a high-quality urban environment, with reduced pollution, lower housing costs – because cars do not need to be catered for – and a safer lifestyle. Cars remain available, however, through neighbouring car clubs. In the summer of 1997, Edinburgh ran a design competition for a car-free area in the west of the city.

Urban Villages
The local authority aims to promote the existing strong local communities which form Edinburgh to allow them to become, as far as possible, self-sufficient. The buzz phrase is 'reinforcing the potential for local sustainability'. The villages are Portobello, Leith, Morningside and so on.

School Travel
In the 1980s, eight out of ten children travelled to school by public transport, now just one in ten get the bus to school. Children are therefore being given the message that the car is vital for short journeys. In Germany and Holland, the proportion of children travelling to school by car is very small, but both countries have higher ownership than the UK. If changes are to be made, a safer environment must be created.

Road Pricing
This could be a key plank of Edinburgh's strategy, and offers some radical solutions to revenue-raising by local authorities. It is generally agreed that motorists do not pay the full cost of using a car – the cost, for example, of polluting emissions affecting asthma, water pollution and so on. While the cost of driving a car declined by 10 per cent between

1974 and 1994, the cost of bus travel rose by 50 per cent, and rail travel by 70 per cent.

Edinburgh is considering a cordon around the city with the aim of using the income from road pricing to benefit an integrated transport system.

According to the city's transport convenor, if Edinburgh raised £60m a year in road pricing and was allowed to keep £20m, it could create the best integrated transport system in Europe. As road-pricing was introduced, park and ride facilities would be improved, and a metro system would be created. Such ideas are already in use on the continent. In Norway, drivers pay to enter the centre of Bergen and Trondheim, not because of the traffic problems, but because of the value of road-pricing as a revenue-raising device. Edinburgh sees this as one of the best ways to cover the externalities which are not taken into account in the running of a car. The city's optimism is based on the belief that the quality of the air we breathe is an important social and political issue, and that the public themselves will eventually see the necessity of such measures. Support can also be found in the Labour Manifesto, which says the party aims to tax activities it wants to discourage. The Government might also choose to invest the money raised through fuel price rises in alternatives to car use. More radically still, personal taxation could be abandoned and switched to much-increased taxes on energy and other resources.

Road Transport and Logistics

Logistics is an off-putting way to describe the flow of goods in and out of industries. At its simplest, it means freight travel. The issue is hugely important to business. A study of European manufacturers found in 1992 that firms spent an average of over 10 per cent of sales revenue on logistics – which also includes storage, materials handling and order processing.

When companies were asked about 'green logistics', the majority said they would expect environment-friendly transport to mean higher costs. Sometimes this is not the case. Professor Alan McKinnon, an expert on freight transport issues at Heriot-Watt University, quotes the example of a German washing-powder manufacturer which reduced its distribution costs by 15 per cent by switching from road to rail, and at the same time cut the exhaust emissions from its road delivery vehicles by 7600 tonnes a year. If the externalities of production and consumption are to be accounted for in future – for example, if power generators were to become responsible for the damage caused by their polluting emissions – such considerations could become very important.

Externalities in the transport sector would include road congestion, oil spillages at distribution centres, road traffic accidents and vibration-damage to roadside buildings.

The Royal Commission on Environmental Pollution estimated in 1994 that the external costs of road freight, from accidents, noise and air pollution, was between £1.6 and £3.6 billion a year. It has been estimated that to make freight traffic pay for its costs, UK lorry taxes would have to increase by around 100 per cent, which would raise transport costs by around 30 per cent.

The need to strike a balance between industry and environmental sustainability is particularly compelling after three decades in which improvements to the Scottish road structure have improved business competitiveness. Today, with increasing road traffic congestion on the M8 and in the Midlands and southern England, it has become obvious to business managers that road freight is not everything.

Other changes in business have contributed to the problem. Materials are now being sourced over much greater distances, and overseas markets have expanded. The proportion of road freight journeys crossing international boundaries increased from 20 per cent in 1986, to 31 per cent in 1991. Average journeys are much longer. The current business practice of contracting out functions previously carried out in a single factory has also meant more freight movement at intermediate stages in the production process.

In a peripheral country like Scotland, companies may

begin to suffer 'transit-time penalties' as congestion increases. The answer, which would also provide environmental benefits, may be to develop a range of traffic strategies, involving some faster rail-freight services via the Channel Tunnel, North Sea ferries and the container routes out of Grangemouth, Teesport and Hull. If direct shipping services and air-freight services were expanded, these would also reduce Scotland's reliance on road movements.

McKinnon suggests that in future the twenty or thirty years of lorry traffic growth may be viewed as an episode comparable to the earlier growth of canal or railway traffic. With the infrastructure now largely in place, and new roads therefore making little difference, the decline in road infrastructure investment has already begun, although the Department of Transport estimated in 1995 that the amount of heavy traffic on British roads would increase by between 56 per cent and 115 per cent between 1994 and 2025.

However, in the UK as a whole, it has been shown that only around 29 per cent of traffic is business-related, and many road freight operators are under-utilising the available road capacity. The situation could be improved by an increase in return loading – typically one vehicle-kilometre in three is run empty – and improvements in load planning. It has also been pointed out that only 9.2 per cent of lorry traffic in 1992 ran overnight.

McKinnon suggests that better use of the existing road capacity is important, and proposes aggressive road pricing as one means of control. The use of 'congestion tolls', for example, would allow business users to pay for easier traffic conditions, while marginal car traffic was diverted, re-scheduled, or switched to public transport. Pricing would also give industry an incentive to manage its distribution systems better.

However, it should be noted that Scotland manages to cope with peripherality partly because it has developed high-value goods: whisky and electronic equipment account for two thirds of our export market.

The Immediate Future

.

Before the election of the Labour Government in 1997, previous administrations were not disposed towards planning for reductions in car travel, although the last Conservative government did set up a national transport forum in February 1997, and produced a Green Paper, *Keep Scotland Moving*, in which it advocated a car-neutral policy. I hope this chapter has shown that a car-neutral policy is not the way ahead.

In England and Wales, Tony Blair's Government immediately put environment and transport under one department umbrella and, in June 1997, announced that most road-building plans in Scotland, including a controversial extension to the M80 in the Kelvin Valley, would be put on hold pending the outcome of a wide-ranging transport review.

Some schemes – the replacement of the Kincardine Bridge over the Forth, and the closing of the gap in the M8 between Baillieston and Shotts – were allowed to proceed, but major projects were mothballed including the 'dualling' of the A1 between Haddington and Dunbar and part of the M77 south of Glasgow. Improvements to the A9 at Helmsdale, the A92 at Balmedie, north of Aberdeen, and the A830 at Arisaig, were also put on hold.

Strong action is unavoidable if the Government is to deliver its optimistic target of a 20 per cent reduction in CO_2 levels by 2010 – transport in Scotland is currently responsible for 32 per cent of CO_2 emissions. Ideally, the Government's White Paper on transport, to be published in 1998, should see a move against road-building and a funding switch to public transport.

In the immediate future, we will see a range of measures. Research indicates that the stick of increasing charges, particularly parking charges, is more successful than the carrot of providing public transport improvements. The same conclusion is supported by a Scottish Office study which found that the toll on the Forth Road Bridge would have to increase from 80p to £6 in order to keep traffic volumes at 1994 levels by the year 2005. Similarly, studies show that limiting CO_2 emissions by fuel price controls alone would mean an 80 per cent increase in the cost of petrol by the year 2000.

The Department of Transport estimated that a combination of car restraint, improved public transport and road pricing could reduce CO_2 emissions by 22 per cent, and make cities more pleasant places. Urban councils may also have to grapple with the fact that more than half the off-street parking spaces in city centres are privately owned and are therefore not controlled. In Edinburgh, 60 per cent of the vehicles in the morning rush hour are heading for a private, non-residential space.

Buses are also set for change in the short term. The first integrated bus system running on North Sea gas will be operational in Stirling by the year 2000. Seven buses will converge on the town centre and castle from park and ride centres, with moving walkways linking the central coach and railway stations. The next generation of vehicles will use natural gas, which will cut nitrous oxide emissions by 80 per cent and particulate emissions by 75 per cent, when compared with diesel engines.

The business of road design and engineering also offers potential for improvement. A new form of tar has been invented which will significantly reduce the polluting run-off from roads. The permeable road surface allows water to trickle through to the aggregate below, which filters out impurities, and may also contain 'superbug bacteria' to consume waste oil. The new surface is safer because it is less slippery, will cut traffic noise by half and has anti-spray properties which will make it attractive to pedestrians. A related design, a 'French drain' which performs the same functions at the roadside, has been installed along the A1 from Tranent to Dunbar. However, by the summer of 1997, no design contractor had put the system forward in any proposed road scheme.

Another forthcoming development is the control of fumes from petrol station pumps. Vacuum pumps have been fitted to the handles of petrol pumps in Sweden for more than a decade, and similar systems will be implemented throughout the EU to capture and recycle the vapours.

Finally, in the current situation, with a deregulated bus network and privatised rail companies, the way ahead must involve partnerships at regional levels, within a national framework.

Because our transport problems north of the Border are not too acute, we have an opportunity to act before it is too late, and, perhaps, before our public transport and rail systems are run down any further. Because we have a relatively low level of car ownership, compared to more developed regions, it should be easier for Scotland to move to a more balanced transport system, with greater use of public transport. We must also see the development of rail and water-borne freight services in Scotland. The Scottish Association for Public Transport (SAPT) has suggested these two sectors should be trebled in size by 2020.

Under that scenario, lorries would still be handling 70 per cent of all freight, but rail and water could deliver 70 per cent of long-distance freight, with a consequent dramatic reduction in traffic on the M74.

SAPT has also proposed the introduction of licences for peak-period access to congested corridors, such as the M8, and to Edinburgh, Glasgow and Aberdeen, with the proceeds going to public transport, and the conversion of the annual increase in petrol taxation to a 'fuel surcharge' to support sustainable transport programmes.

Another stage of reform has been suggested in which a Scottish Parliament might devolve powers to a series of regional transport and land use planning boards.

In a speech in 1997, David Begg summed up the current situation, and proved Mrs Balfour, and Queen Victoria, quite correct. He said: 'The sharp rise in atmospheric air pollution, increasing congestion, and the developing mobility deprivation of those without access to cars makes it imperative, and not just desirable, that policies aimed at reducing car dependency and encouraging public transport, walking and cycling are implemented as quickly as possible.'

The Future

·

Research has indicated that the most important influence on the level of CO_2 emissions over the next thirty years will be the rate at which traffic volumes grow, not the cleanliness of individual vehicles. None of the impressive advances in fuel efficiency, engine technology and vehicle design will reduce the number of vehicles on the road.

But the ultimate in clean, green cars – the hypercar – is already being developed. The concept was developed by Amory Lovins, of the Rocky Mountain Institute, a building which stands at an altitude of 7000ft, but needs no heating or cooling equipment.

Lovins, once voted by the *Wall Street Journal* as one of the people most likely to change world industry, set up a think tank which showed that by combining ultra-light, ultra-slippery carbon fibre with hybrid-electric propulsion (a mix of fuel and battery systems), it was possible to improve fuel efficiency between 400 and 1000 per cent, and to reduce pollution 1000-fold.

This would allow you to drive from Edinburgh to London on less than a gallon of petrol in a car which would be stronger, more comfortable and faster than anything today. Lovins did not patent the idea, but floated it to the vehicle industry to encourage competition. At least twelve manufacturers, including Honda, are now working on the concept. The cars might also be covered in solar cells, to generate electricity while they were standing still. When Lovins aired the idea in 1994 he was told it was impossible, but the first models could be purring by the turn of the millennium.

Hypercars may be the intermediate future, but I believe that in the centuries to come, people will see the last years of the second millennium as a period of simplistic and crude transport systems, a messy free-for-all and a testament to the failure of governments, planners and local authorities. It may also be seen as a time of primitive machines which were fantastically energy-inefficient and used prodigious amounts of precious fossil fuel. In the years ahead, oil is far more likely to be used to create important new materials. Burning finite and valuable natural resources in an engine will appear laughable.

It is hard to imagine the vehicle of the future, but cars, if they exist in a recognisable form, could be comfortable capsules, without a driver's seat or steering wheel, programmed to travel along electronically-controlled highways, above or below ground. It may be difficult to tell the difference between buses and trains, if, indeed, there is any difference. But there will be some means of mass transport, which will be comfortable, air-conditioned and fast.

It is even possible that we will have developed a society in which the need to travel on a daily basis is much reduced.

10

·

Cities and Sustainability

·

The state of society now leads so much to great accumulations of humanity that we cannot wonder if it ferment and reek like a compost dunghill. Nature intended that population should be diffused over the soil in proportion to its extent. We have accumulated in huge cities and smothering manufactures the numbers which should spread over the face of a country and what wonder that they should be corrupted?

Sir Walter Scott, *Journal*, 1828

AROUND 80 PER CENT of Scotland's 5.1 million people live in towns and cities which cover just 3 per cent of the land area. Scotland has one of the lowest population densities in Europe, but sits next to England, which has one of the highest. In the sparsely populated Highlands, where numbers are still dropping, the 300 000 inhabitants are vastly outnumbered by 2.5m sheep.

The population centres are concentrated in the Central Belt and grew up around heavy and polluting industries including coal, steel and shipbuilding. It is in towns and cities that the majority of us have our daily experience of 'the environment', and it is in towns and cities that the battle for an improved environment will be won or lost.

Although most of us still think of nature as a countryside issue, the biggest concerns of the day are urban. The global warming caused by CO_2 emissions is an urban issue, so is transport, and so is energy use and efficiency. And it is in urban areas that biodiversity has been overwhelmingly diminished. In parts of the grey, urban world, greenness may be found only in public parks, back gardens, road and rail corridors, canals and graveyards. But where large areas of green space are preserved, such as Arthur's Seat in Edinburgh, and Pollok Country Park in Glasgow, they act as welcome lungs for the city and its inhabitants.

A major task facing planners, government and city councils today is to improve cities as they are, without forever expanding their boundaries. In an attempt to stop the spread, greenbelts were introduced from 1955, but have too often been damaged by the wrong planning decisions. The six Scottish greenbelts – Glasgow, Ayr/Prestwick, Falkirk/Grangemouth, Edinburgh, Aberdeen, and Bridge of Allan and Ochils – are continually threatened by out-of-town shopping and business centres, and new housing and industry.

At the Earth Summit in Rio, it was agreed that environment and development conflicts should be reconciled. That does not mean new hypermarkets on the edge of the motorway. But there is something to be said for the fact that we are talking about sustainability and that councils are considering how they can implement Rio's Agenda 21, which was intended as the blueprint for city living in the twenty-first century. Improving our cities means dealing with the five issues which define cities everywhere: poverty; homelessness; transport congestion; waste disposal; and energy use.

Professor Michael Carley, of the school of planning at

Heriot-Watt University, suggested the key to sustainable development in the 'city-region' of the twenty-first century was 'integration between economic vitality and quality of life; between the needs of the neighbourhood, the city and the region; between long-term sustainability and short-term politics; between "bottom-up" local innovation and "top-down" national urban policy.' He also said decision-making must mean empowerment, and the involvement of local communities. Tackling the alienation of people in the street, and the housing estates, is a key theme of Agenda 21. In a paper delivered at a Glasgow conference in June 1996, the professor suggested empowerment meant an end to officials and elected councillors automatically thinking that they knew best, or pursuing a narrow political agenda.

On a similar theme, a report from Friends of the Earth said that a sustainable city was one which met the needs of its inhabitants and provided a good environment, without damaging other areas as a result. It added that cities had to maintain and improve the quality of life of their residents, while reducing CO_2 emissions and the use of non-renewable resources, without expanding water consumption.

Without exception, modern Scottish cities exhibit social, economic and environmental problems, including isolation, the breakdown of communities, the inaccessibility of services, poverty and unemployment, air pollution caused by traffic, and waste. Urban regeneration, therefore, is an essential part of city development, and we cannot seriously claim to have sustainable cities while they suffer from bad housing and crime.

There have been numerous regeneration schemes and community projects over the years, including community enterprise and small business initiatives, but none has been able to deal with the underlying problems of unemployment and poverty, which should be regarded as important by all of us. Creating a society of more equal opportunity will mean ditching the trickle-down theory which states that the creation of wealth benefits everyone in the end. In fact, the number of millionaires in Britain has been rising at the same rate as the number of homeless, according to Linda Dunion, the former director of Scottish Education and Action for Development (SEAD).

Dealing with poverty will require a radical shift in policy making. For example, in 1996, Ms Gray quoted an initiative from Sri Lanka, where low electricity use is rewarded by a low tariff. In this country, the biggest business consumers pay least. FoE calculated that 17 000 jobs could be created by a programme of energy conservation, and more through the expansion of public transport. If such programmes were accepted, it might well be possible to demonstrate how social, environmental, economic and moral imperatives are linked.

Ms Gray also quoted the example of a community action group, the Kirkwood Food Co-op in Coatbridge, which was set up to deal with the lack of access to fresh food in an area with few good shops, a poor transport infrastructure and a large number of people on low incomes. The organisation provided job placements for young people, assumed a health education role, and sold hundreds of pounds' worth of fresh fruit and vegetables each day.

Gus Speth, the administrator of the UN Human Resources Development Programme, summed up the above points quite neatly. He said: 'Sustainable human development doesn't merely generate growth, but distributes its benefits equitably; it regenerates the environment rather than destroying it; it empowers people rather than marginalising them; it enlarges their choices and opportunities and provides for people's participation in decisions affecting their lives. Sustainable human development is development that is pro-poor, pro-nature, pro-jobs, pro-women. It stresses growth with employment, growth with environment, growth with empowerment, growth with equity.'

Superficially, Scottish cities appear to be closer to sustainability than some English cities due to their lower levels of car ownership and greater use of public transport, but these statistics are simply reflections of a lower standard of living. Unemployment is much higher than it should be, due in part to the decline in traditional industry. The goal of sustainability was put back by the weak planning years of the

1980s, which saw booming decentralisation and the creation of peripheral housing and retailing estates which demand access to a car. Such developments not only add to congestion and air pollution, but marginalise people who do not have cars. It is likely that the sustainable city will be made up of multiple-use neighbourhoods of shops, homes, workplaces and leisure facilities, probably with a wide mix of income levels among residents. Some progress could quickly be made by giving people an opportunity to participate in the planning process, by allowing them to become involved in, and to comment on, housing reviews, environmental assesssment and sites for recreation and industry. Current participation rates in planning matters in the UK are around 5 per cent, compared to 70 per cent in other northern European countries.

It is also the case that most of our neighbours, including Spain, France, Germany and the Netherlands, exhibit a trend towards regional government, while local government reorganisation in Scotland in 1995 resulted in regional planning being dismantled. There are strong reasons for arguing that regional planning is a necessary element of sustainable development, as cities cannot be divorced from their hinterland. In the case of Glasgow and Edinburgh, it is common for people to commute from up to 35 miles on a daily basis – sometimes through two or three planning districts! Voluntary regional planning arrangements have been set up involving groups of unitary authorities, but these bodies are unlikely to take tough decisions on transport, land use and energy.

These issues are particularly relevant in Glasgow, which has become the second biggest centre for shopping in Britain. Professor Carley has warned that if we do not rescue regional planning – for example, through the Scottish Parliament designating a small number of statutory, non-voluntary planning agencies to carry out regional planning – the cause of sustainable development will be set back.

Poor planning has, in the past, resulted in the separation of land uses, the fragmentation of services, goods and accommodation, and the creation of soulless housing estates. New housing needs to be contained within existing city boundaries where possible, with a particular emphasis on the so-called brownfield sites which have been used for industry or housing in the past, rather than the precious greenfield sites which seek to stop the spread.

The organisation of our cities, our transport networks and the way our homes are heated provide considerable scope for improvement. Cities produce large quantities of waste which is dumped in landfill sites, and they use power inefficiently. With good building design and the use of solar energy, our homes and office blocks could capture heat much more economically. Where modern buildings are properly designed, they should require little or no external energy input. Surprisingly, our tenement blocks are a bonus in environmental terms, offering the potential of high-quality homes in energy-efficient buildings, at high population densities. Tenement living makes public transport more viable, and offers the option of district or group heating schemes.

Finally, Professor Carley believes that action on sustainable development will require all the agencies and authorities with an interest in urban issues to concentrate on employment opportunities. 'The time has come,' he said, 'to stop treating public-sector tenants and users of public transport as second-class citizens compared with home-owners and drivers of swish company cars.'

There are no easy answers, but some commentators have suggested that eco-taxes on polluting industries could raise millions of pounds for local economic development, training and urban regeneration. Meanwhile, the obvious conclusion to draw on the current state of thinking on sustainable development is that there has been a strong growth in awareness of the issues, but the bridge between social, environmental and economic concerns has not been made. It is possible that a council roads department may be thinking very seriously about how to deal with the issues raised by transport congestion and pollution, while the personnel department – which could ask people to cycle rather than drive – fails to recognise any role for itself.

Sustainable development in cities is a business for planners, architects, energy managers and educationalists. For everyone.

Some Issues

·

Solid Wastes

Every household produces around one tonne of waste a year, including high quantities of non-biodegradable plastic. But despite official goals of minimisation, recycling and recovery, Scottish local authorities only recycle around 3.8 per cent of all domestic waste. Dundee, which deals with 18 per cent of the waste stream, is one of the few exceptions. There is little hope of cities attaining the official target of a 25 per cent reduction in domestic waste levels by the year 2000. At the end of 1996, fewer than one in six Scottish councils offered plastic recycling banks, compared to a national figure of one in three.

And while advances have been made in the past ten years, the thinking may still be muddled. For example, while every kilo of glass recycled saves 4.8 mega-Joules of energy, each mile travelled to a bottle bank by car consumes 5 mega-Joules. The best option for recycling schemes is kerbside collection, but they are more expensive than asking people to burn petrol to visit 'bank' sites.

In 1994, a total of just over 16 million tonnes of controlled waste (half of it from construction and demolition) was treated and disposed of in Scotland, 92 per cent of which went to 500 landfill sites. Industry accounted for 70 per cent, and households for 15 per cent (2.41 million tonnes). Just 850 000 tonnes were collected for recycling, and between 1989 and 1994 the amount of waste produced by industry was said to have increased by a remarkable 75 per cent.

Around 60 per cent of landfill sites – giant holes in the ground lined with plastic – are leaking highly polluting leachate. They also produce large quantities of carbon dioxide and methane gases, and are sources of litter and bad smells. UK sites emit two million tonnes of methane a year, or around 48 per cent of all emissions, making landfill the biggest source of man-made methane additions to global warming. Less than 5 per cent of sites have the technology to utilise the gas in energy production.

All forms of waste treatment have environmental costs: incineration emits CO_2, SO_2, NO_x and methane (CH_4), and produces ash which has to be disposed of.

In 1996, household waste alone included nearly five billion plastic bottles. In Switzerland, 70 per cent of bottles are recycled, and Norway and Sweden use returnable plastic Coke bottles which can be refilled twenty-five times. In Scotland, the development of recycling of all types is hampered by the low cost of landfill, particularly in the central belt. In rural areas, especially the Highlands, serious problems have been caused by tighter regulations which have made many small municipal landfill sites illegal. Plans for a 'super pit' near Inverness have, understandably, run into the Nimby (not in my back yard) syndrome. Unfortunately, more emphasis is being placed on finding the right site, than on reducing waste flows. Meanwhile, local authorities are braced for a European directive on landfill which demands much higher standards from our sites, and requires significant reductions in the waste stream.

Waste is the responsibility of SEPA, which produced a draft national strategy in 1997. It advocated the 'closed loop' approach and stated that the principle of equity required each generation to take responsibility for its own waste. Its governing principles included clean production processes, the marketing of products which minimise the effect of production, and the extraction of secondary resources or energy from waste.

It also listed a range of waste management targets which have been proposed in recent years by government and the voluntary sector. They include:

1. Reducing the waste going to landfill by 60 per cent by 2005.
2. Recycling or composting 25 per cent of household waste by 2005.
3. Accessible recycling facilities for 80 per cent of households by 2000.

4. Waste management systems for 50 per cent of companies with over 200 employees by 1999.
5. Recovery of 95 per cent of old vehicles by 2015 (the current figure is 75 per cent).
6. Recycling of 58 per cent of waste glass by 2000.

Under 'producer responsibility', legislation already requires companies which produce large amounts of packaging to recover 52 per cent of all materials by 2001, and the newsprint industry has agreed to use 40 per cent recycled paper by the year 2000, an increase of 10 per cent since 1994.

Meanwhile, the landfill taxes of £2 a tonne for construction waste and £7 for more dangerous wastes, which were introduced in 1996, have already encouraged increased reuse of waste materials. There are all sorts of mechanisms, many of them market procedures, to reduce waste. Local authorities have the power to pay third parties a credit for each tonne of waste they divert from the normal waste stream. In 1994, around 30 per cent of councils were using the powers to some extent. Yet we are – in familiar fashion – behind many other European nations. Italy, for example, brought in a tax on plastic bags which resulted in an immediate drop in use of 40 per cent. It also introduced a tax on farmers with more than 200 pigs and no waste facilities!

The concept of zero emissions from manufacturing takes the philosophy of clean technology to its conclusion. At present, it is no more than a concept, but it is being studied around the world. It foresees a network of companies feeding off each other's by-products in the way that organisms exploit different niches in a natural ecosystem. The philosophy behind zero emissions is called industrial ecology. In natural ecosystems there is a closed loop, and no waste. In industrial systems, materials move in a linear fashion from manufacture to consumer and to waste. A shift in thinking is required in order that products are not seen as the end of the line, but, as SEPA's waste strategy says, 'a temporary embodiment of materials'. Research into zero emissions is being carried out in Tokyo.

Sewage sludge

Sewage sludge dumping at sea will be banned in 1998, and two key options for dealing with the waste have been developed – incineration and disposal on land. At present, 67 per cent of the sludge is dumped at sea, 20 per cent is used on land, 10 per cent goes to landfill, and 3 per cent to incinerators.

The former Strathclyde Region, which has since been replaced under local government organisation by unitary councils, decided to use sludge and sludge pellets, which contain nitrogen and phosphorus, as a fertiliser for agriculture and forestry. Trials on forestry plantations in the north of Scotland have proved successful. It might also be used in the process of reclaiming derelict land.

The old Lothian Region decided to incinerate its waste in a new, high-technology plant which will generate 21MW of electricity. The ash left at the end of the process will go to landfill. In strict environmental terms, disposal to land is more attractive, although assessments must be carried out to establish that the land is suitable, and that the sludge will not increase phosphorus levels in water courses. In Germany, both methods have proved controversial because of the polychlorinated biphenyls and dioxins which can be deposited in the soil, or released to the air.

Energy

Cities are responsible for around 75 per cent of national energy consumption, and waste large quantities through poor housing stock. However, no details are available from Scottish utilities on how much power individual cities use, and how much CO_2 they are responsible for. In Newcastle, investment in energy efficiency measures has reduced energy consumption in municipal buildings by over 50 per cent. The city has also been introducing small-scale 'combined heat and power' schemes and improving insulation standards in its housing stock. Still, even in a city which is making significant strides, only 4 per cent of the energy used comes from renewable resources. The council aims to increase the figure to at least 10 per cent in the next 10–20 years.

One statistic from Glasgow illustrates the problem of fuel poverty and energy waste in the domestic sector. The city is currently dealing with around 30 000 'Wilson block houses', most of which are council stock, in which thermal imaging showed that the curtains were providing more insulation than the walls. Glasgow City Council has set itself a target of eradicating fuel poverty in 80 000 homes over five years. But it remains unclear how it will achieve the goal.

Vacant and Derelict Land
Scotland's industrial past has left, at the last count, 14 100 hectares of vacant and derelict land at 6000 public and private sector sites. The sites may be in areas which are ideal for development, but the extent of the contamination, and the causes, are often unknown. As a result, developers prefer to opt for greenfield sites. In theory, the polluter is responsible for cleaning up contaminated land, but where sites are derelict the guilty company may no longer exist. In the past, sites have usually been 'treated' by removing the soil to a landfill facility. Attempts are now being made to deal with contamination on-site.

Institutional Problems
Different agencies have responsibility for different issues and areas of development, and even individual departments within councils may not adopt the same attitudes. For example, the roads department and the social work department in Glasgow City Council may not be equally convinced of the merits of sustainable development. And if Scottish Enterprise and its inward investment arm, Locate in Scotland (LiS), are not operating on the basis of sustainability, then the actions of concerned local authorities are diminished. LiS, in fact, issued a letter to all councils in 1996 asking them each to nominate two greenfield sites for inward investment and the development of factories. They stated that the sites had to be near to motorways, and thereby immediately failed a key test of sustainability. The

fragility of the agency's policies were demonstrated by the delays to the massive Hyundai electronics factory on farmland outside Dunfermline. The project was meant to bring 2000 jobs, but was halted by the collapse of financial markets in the Far East. It is a pity that Scottish Enterprise has not promoted indigenous industry with the same fervour that it has pursued inward investment over the years.

The principles of sustainable development have also been lacking in the actions of NHS trusts which have been selling inner city sites to pay for 'greenfield' developments which increase traffic densities. And, with hindsight, we can see that planning decisions in the 1960s, which directed industrial development to new towns, left Glasgow with rising unemployment and large areas of derelict land.

Environment organisations have been quick to point out that the official agencies which are in the best position to deal with urban environmental problems – Scottish Enterprise, LiS and the Local Enterprise Companies – seem least switched on to modern enviro-economic thinking.

Litter

Litter is a problem almost everywhere in Scotland because of public ignorance. The materials involved are invariably durable plastics, wrappings and aluminium, none of which readily biodegrade. Some of the most affected areas are towns and cities and the road and rail routes connecting them, but litter is also a problem on seashores, where it is deposited by people, sewage and shipping.

Radioactivity

This is a complex subject dealt with by the media in a simplistic way. We are all exposed to radiation from natural and man-made sources. But few of us are aware of the fact that around 85 per cent of the annual dose received by each individual comes from the rocks around us, while 14 of the remaining 15 per cent comes from medical sources, such as X-rays and scans. Just 1 per cent comes from fallout, or the discharges from nuclear sites. The balancing statement is that nuclear plants have to contain radioactivity which, if it escaped, as it did at Chernobyl, would have disastrous effects over very large areas.

Natural sources of radiation include cosmic rays, gamma rays from the earth, radon products, and natural radionuclides in our diets. Radon comes from uranium occurring naturally in the ground and is readily dispersed in the atmosphere. However, it can become a threat if it builds up inside a house. Indoor levels depend on the prevailing geology, atmospheric conditions and the degree of ventilation. Generally speaking, homes in the Central Belt have low radon levels, while parts of Caithness, Sutherland, Kincardine and Deeside have concentrations above 'action levels'.

The atmospheric testing of nuclear weapons in the 1940s and 1960s resulted in annual doses of around 140 microsieverts. The figure has dropped since the implementation, twenty years ago, of the partial nuclear test ban. The international limit for exposure to man-made radioactivity is one millisievert a year.

The Chernobyl disaster in April 1986, which caused hundreds of deaths from cancers in Belarus and Ukraine, caused a fivefold increase in the annual UK dose due to fallout. Southern Scotland suffered significant deposition of radiocaesium in the rainfall which followed the accident, although the levels remained below international safety standards. However, in 1997, thirty-six farms and 76 000 sheep were still subject to market restrictions because of the incident.

Scotland has twenty-seven monitoring stations designed to detect an overseas nuclear accident, and supplementary checks are carried out on food and drinking water. Scottish nuclear power plants have strict guideliness and assess their discharges by checking their workers. The discharges have consistently been within authorised limits. However, relatively high levels of caesium-137 were found in coastal waters off west and north Scotland during the 1970s and 1980s, due to emissions from the Sellafield reprocessing plant in Cumbria.

Transport

This is one of the major issues for planners pursuing sustainable development. The solutions to our current problems are likely to include: the opening and reopening of rail lines; prioritising bus lanes; better park-and-ride facilities; more home deliveries by retailers; increased return-loading for lorries; and the development of urban villages in which many journeys are possible on foot. The Edinburgh Conference Centre, a partnership project which involved the public and private sectors, is a good example of an important development accessible on foot from major hotels.

Planning

Planners and developers must consider how people should travel to new buildings, the energy consumed by the building, and the requirement for raw or secondary materials. Developments should be contained, where possible, on brownfield and derelict sites, and not on greenfield sites. Buildings should be built, as far as possible, with local materials and recycled materials. They should be appropriate to the local climate, and should be designed to last for at least 100 years. A new culture is also required within the building trade, so that existing regulations are treated as minimum standards to be exceeded, and not high standards to be grudgingly met. These are not revolutionary ideas. In Germany, the Netherlands and Switzerland, building practices commonly exceed the minimum standards asked for in law.

Sustainable Glasgow?

Glasgow City Council is aware that moving towards sustainability is not an optional extra. Some senior officials even accept that the development of a strategy which gives equal consideration to environmental, social and economic issues could lead to great improvements in the city without incurring additional costs. Where councils are reluctant to accept such radical thinking, they may be required to act by European and national legislation.

The local authority's policy objectives, laid out in a discussion paper, included:

1. The promotion of environmental education.
2. The regeneration of the economy through the reuse of vacant and derelict land.
3. Tackling inequalities, the regeneration of communities and the promotion of health.
4. Monitoring and reporting on the physical environment.
5. Reducing waste, energy use and the impact of transport on the environment.

The authority's immediate proposals, for 1997–8, included more public participation in decision-making, and a new dialogue with community and single-interest groups. It proposed anti-poverty policies, for example, which recognised that housing and health were 'closely associated with environmental issues'. The discussion paper accepted that improving the quality of the built environment could reduce insecurity, stress, vandalism and general decline. It identified energy as an issue of great importance, locally and globally, and recognised that CO_2 emissions could best be tackled at the city level.

The authority aims to build on the regional council's Sustainability Indicators report, published in 1995, which helped to illustrate the scale of the problems facing every British city. In the section devoted to the physical environment, the indicators included freshwater quality, air quality, land use and woodland cover. It accepted that air quality was of 'major and increasing' concern, and that poor air quality had been linked to a variety of respiratory diseases. It showed that nitrogen dioxide levels between 1993 and 1994 in Hope Street, in the city centre, were close to the EU limit, and broke it on several occasions. It also revealed that the amount of vacant and derelict land had increased substantially since 1980, but that brownfield sites were being used more often for new housing than they were in the 1980s.

Car ownership was shown to be at a relatively low level of 258 vehicles per 1000 people in 1993, although traffic volumes in Glasgow had increased by 67 per cent in the ten years from 1984. They are currently increasing at around 4 per cent per annum.

The new city council will prepare its own sustainability indicators, but while it grapples with a financial crisis which has led to school closures and cuts in services, it is unlikely that great changes will be made. Glasgow, like many other cities, understands that action needs to be taken on environmental issues, but finds it hard to translate policy documents into action. It has produced a City Centre Millennium Plan which aims to reduce traffic by 30 per cent, but has no wide-ranging proposals in force to deliver the target.

Most of the 850 000 tonnes of waste which the city generates each year is still sent to landfill sites. The city presently has no hope of achieving the government target of recycling 25 per cent of household waste by the year 2000. In 1992, the regional council recycled less than 4 per cent of the area's waste.

Sustainability

Sustainable development is difficult to define. (John Major, 1994)

The issue of sustainability – reconciling the conflicting interests of the environment and development – underpins this book. In the last section of this chapter, I want to draw on a discussion paper produced by Friends of the Earth (FoE) Scotland, *Towards a Sustainable Scotland*, and, to a lesser extent, on the former Conservative government's UK strategy for sustainable development.

The environment group's sister organisation in the Netherlands has studied the idea of 'environmental space' in some detail. According to FoE's Scottish report, its attraction is that it moves sustainability out of the arena of conservation, into the social and economic mainstream. It also teaches us that sustainability cannot be attained by one country alone, or achieved inside international boundaries.

For example, the UK could only become self-sufficient in wood products if it covered 94 per cent of the country in plantations. Therefore, to live within our own environmental space, we should resolve to expand production and cut consumption. Our 'space' is defined as the amount of a resource that can be consumed without damaging the capacity of the planet to support us, and every other species. In the case of wood products, our environmental space would include the Scandinavian forests from which we import timber.

The government strategy document states specific principles, including decision making based on the best scientific advice and analysis of the risks, the use of the precautionary principle, and a consideration of the ecological impacts of development, particularly where resources are non-renewable. It also says that the polluter should pay.

The key principles outlined in the FoE document are:

1. The precautionary principle – which means not confusing 'no evidence of damage', with 'evidence of no damage'.

2. The equity principle – which means industrialised nations should cut consumption, while poor nations consume more. Equity also means China has the biggest claim on world resources.

3. The proximity principle – which means not exporting problems: that is, not siting polluting industry in the Third World to avoid environmental regulations, and not exporting waste. It also means using European, rather than tropical, hardwoods.

If we considered the role of nuclear power in a sustainable energy policy for Scotland, we would find that it breached all three principles. It does not satisfy the precautionary principle because although the risks of an accident are small, the consequences are enormous. The equity principle is violated because we would leave future generations with a legacy of radioactivity and few of the benefits of the energy generated. And it would breach the proximity principle because of the costs of uranium mining in other countries, and the import of waste for reprocessing in this country.

Putting sustainability into practice, according to the

government document, will involve all sectors of the community, from central government to individuals. It will involve international agreements, government policy making based around clear objectives and targets, and a recognition from business that good environmental performance can increase competitiveness.

The strategy also recognised that the costs of environmental damage had to be built into the prices charged for goods and services, a proposal which demands the development of environmental accounting. It suggested sustainability indicators – like those drawn up in Glasgow – would be required to judge progress.

Finally, using the FoE study, I want to look at the issue of sustainability in terms of three important issues for Scotland – non-renewable materials, forestry and industry and services.

Non-renewable Materials

In terms of environmental space, the limiting factor for non-renewable resources, like bauxite or aluminium, is the environmental damage caused by extraction. The Wuppertal Institute in Germany has recommended a cut in global consumption of primary non-renewables of 50 per cent by the middle of the next century. When global disparities are taken into account, this means a cut of 70 per cent in Scotland, with an initial target of 25 per cent by 2010. This would require a massive switch to secondary resources, such as recycled material, and a process of de-materialisation – reducing the demand for raw material.

The final calculation involves not just the aluminium at the factory gate, but the rock and soil removed in the process of extracting the bauxite. Four tonnes of waste are produced for every tonne of finished aluminium. Most of the world's bauxite comes from surface mines, and so the process also removes vast quantities of topsoil.

Aluminium production alone is said to account for 1 per cent of global energy demand, while cement manufacture – from the raw material lime – accounts for a remarkable one third of world energy production. European studies have suggested a 70 per cent reduction in the UK consumption of cement by 2050.

Invariably, it is the poor nations which supply the raw materials demanded in the developed world, and which bear the brunt of environmental degradation. Developed countries have either exhausted their native reserves, or want to avoid the social and environmental costs of extraction.

The Worldwatch Institute in Washington has calculated that the average American accounts in his lifetime for 540 tonnes of construction materials, 18 tonnes of paper, 23 tonnes of wood, 16 tonnes of metals and 32 tonnes of organic chemicals – or twice as much as the average western European. This infringes the equity principle.

There are no detailed figures for resource use in Scotland, but in general terms we export coal and petroleum products, and import most other non-renewables. Therefore, raising the proportion of the resource coming from secondary materials in Scotland would have a positive effect on the balance of payments. And if nothing is done to reduce the exploitation of primary resources, then reserves will become increasingly remote and difficult to work, necessitating greater energy use and a bigger transport infrastructure.

Current trends towards the more efficient use of resources include the replacement of some metals by ceramics and plastics, and the use of computers in improving resource use efficiency. There is a thermodynamic limit, however, beyond which efficiencies cannot go.

The answer – dematerialising – would include a reduction in fossil fuel use, cutting the demand for land and travel, and altering production and consumption habits to reduce waste.

Products might also be made in ways which allowed them to be dismantled easily to promote reuse. As long as the drinks manufacturer is not responsible for disposing of his end product, there is no incentive to deal with empty drinks containers.

FoE would like to see 50 per cent of waste being recycled

by 2010, VAT being reduced on products containing 100 per cent recycled or resued components, and taxes on non-renewable resources.

Forestry

Forestry is a continental resource. The report suggests that its environmental space should be defined as the European supply, minus 10 per cent for conservation purposes. When the resource is considered against the European population, the environmental space for wood is 0.56 cubic metres per person, compared to the 0.73 cubic metres consumed in Scotland. That would mean a 23 per cent reduction in timber use in Scotland. Our management of forestry and use of timber is also an obvious candidate for change while the UK imports 87 per cent of its timber requirements.

More woodland in Scotland could provide aesthetic, recreational and economic benefits, and an increase of forest cover to 30 per cent is suggested by the study. At present, around 15 per cent of Scotland is wooded, compared to the EU average of 25 per cent. Forest cover figures elsewhere are: Finland 76 per cent; Japan 67 per cent; Portugal 32 per cent; USA 31 per cent; Germany 30 per cent; and England 7.5 per cent.

It has been estimated that for Scotland to grow all the wood it consumes would mean forestry covering 28 per cent of the land area, which is theoretically possible. However, as already indicated, it would be impossible for the UK as a whole to become self-sufficient.

Timber enters the market as roundwood or thinnings; Scotland produces half the UK's output and has one of the four paper and board mills – Caledonian Paper plc, at Irvine – which rely on domestic supplies. There are 100 mills based on imports.

Sawmills import roundwood, paper mills import pulp, and panel and board mills import wood residues. There is a great deal of waste in the system, and there is scope for the use of thinnings and whole trees from the management of urban woodlands. These normally go to landfill sites. Wood waste from furniture making could go to paper and board manufacturers, and the construction industry also generates large amounts of surplus wood. There is an overwhelming case for reducing the amount of packaging, junk mail and office paper waste.

Industry and Services

Industry in Scotland used to be based on easily accessible resources and exports to a global empire. Following the restructuring of recent decades, there are more people employed in service industries than manufacturing. The service sector accounts for over 50 per cent of GDP, manufacturing accounting for 23 per cent, transport for 9 per cent, and tourism for 5 per cent. A fifth of the country's GDP comes from its exports of £8.8 billion, over 60 per cent of which go to the EU, particularly Germany and France.

Ideally, we should reward companies which use resources efficiently and do not cause pollution. Companies need to be valued for providing employment, and for their environmental footprint, as well as their profit margins. Clean operations can also make good economic sense. Japan's tough vehicle emission standards meant that its car makers did not have to alter their production plants when higher standards were introduced. The UK car plants were faced with hefty bills.

Modern industries must aim to reduce resource use, energy consumption and waste production. Businesses could opt to conduct a waste-minimisation analysis of production processes, design products to consume the minimal amount of energy and resources, design products to be multi-purpose, redesign products to increase their lifespan, and use materials from secondary sources.

Porsche, for example, concluded in the 1970s that extending the life expectancy of its cars to between 18 and 25 years would cut resource consumption by 55 per cent. Kodak and IBM have implemented comprehensive waste reduction strategies at their Scottish plants, and the BP complex at Grangemouth is supplying heat to nearby homes. Polaroid at Dumbarton, which produces cameras, films and sunglasses, instituted a 50 per cent waste minimisation plan by dividing its waste into three main categories: hazardous chemicals and ozone depleters; acids, oils and inks; and rubble, rubbish and trash. It also switched to less damaging chemicals, eliminated polystyrene, began reusing chemicals and designed reusable boxes for cameras. When the whole plant was audited, including the plastic cups in the canteen, the company managed to reduce the number of skips it was using each month from twenty-six to twelve. It saved £400 000 in the process. Polaroid is aware of the commercial value of a green brand image.

In the long term, it is likely that companies using energy and raw materials will have to pay the full cost of generation and production, and will have to accept full liability for packaging. In the current climate, government policy concentrates on recycling as an end result, rather than tackling the demand for energy, water and materials during the manufacturing process. Waste is seen as something to throw away, and most businesses are based on short-term thinking and the assumption that processes are already efficient.

The FoE report suggests resource consumption information should be collated annually, and that local databases should be set up to provide small companies with information on waste minimisation.

A Final Thought

In the 1950s, London and New York were the two biggest cities on the planet, each with eight million inhabitants. By 2015, there will be thirty-three megacities with more than 20 million inhabitants. Twenty-one of them will be in Asia. There will also be 100 cities with more than five million people which currently have populations of just 200 000. Eighty of these will be in the developing world.

11

·

The Way Ahead

·

Is there not the Earth itself, its forests and waters, above and
below the surface?

These are the inheritance of the human race ...

What rights, and under what conditions, a person shall be
allowed to exercise over any portion of this common inheritance cannot be left undecided.

No function of government is less optional than the regulation of these things, or more completely involved in the idea
of a civilised society.

John Stuart Mill, *Principles of Political Economy*

SUSTAINABLE DEVELOPMENT HAS BEEN defined as 'meeting
the needs of the present without compromising the ability
of future generations to meet their own needs'. In simpler
language, it means not screwing up tomorrow by what we do
today. When we consider the environmental state we are in,
and the condition of this fragile land, there are a number of
environmental truths we should bear in mind.

First, we have been here before. In recent decades we
have taken action to clean up our air and our rivers, and
succeeded. On a global scale, governments have taken unprecedented, co-operative action to tackle the issue of ozone
depletion, and we understand the mechanisms necessary to
deal with one of the great environmental challenges man
has visited on himself, global warming.

We do not face insurmountable problems. Equally, we
cannot simply blame our difficulties on evil multinational
companies, greedy politicians and profligate dictatorships.
We are businessmen, we are politicians and we are factory
workers. People run the corporations and the industries

which are blamed by environmental groups for all the world's ills.

It would not do, of course, for Greenpeace to attack the man in the street. It would be a risky strategy indeed to accuse their own members of raping and degrading the earth. BP, on the other hand, is a big, disembodied corporation, a handy demon to point a finger at. But Greenpeace members drive cars and catch buses fuelled with the oil won by petroleum companies, and use plastics and consume electricity. As long as we sit back and wait for the environment organisations to win the battle for us and force governments into action, nothing much will change.

Obviously, there are key roles to play for industry and government in the creation of a sustainable society and, more specifically, for the decision-makers in society, the people at the top of these institutions, who also drive cars, have homes, and are concerned about the world they will leave for their grandchildren. We are all culpable, as consumers and producers.

We all crave comfort, luxury and wealth (spiritual or material), and most of us demand a technologically sophisticated lifestyle. That does not mean we must face a future of self-sacrifice, but it suggests that we need to make informed choices about heating and lighting our homes; about how and when we travel; about how we deal with waste and litter; about the food and packaging we buy. Ultimately, it is individuals who will make the difference. Government, after all, is extraordinarily sensitive to popular opinion.

There is, at present, no common will for change, but there is a commonly voiced hope that education will change attitudes in society and produce future generations which are more environmentally aware. The environmentalism of my generation is in its infancy. The green conscience of the individual has its roots in the 1960s; earlier problems such as the smogs of the 1950s were regarded as single issues, and not part of a greater malaise. Environmentalism did not impinge significantly on the national conscience until the 1980s, when a series of environmental incidents, such as the disasters at Bhopal, Seveso and Chernobyl, and the acceptance by world governments of acid rain and global warming as international problems, put an official seal on what the environmental activists already knew: modern lifestyles were unsustainable, and we were producing waste and pollution at a level which the small blue planet could not accept.

That we should have come to such conclusions only recently is hardly surprising. People living in the 1990s – whether grandparents or primary schoolchildren – have witnessed remarkable strides in technology and communication. In my own eighteen-year career in journalism, I have experienced the old and the new worlds. As a trainee reporter on the *Press and Journal* in Aberdeen in 1979, the tools of my trade were a pen, a notebook, a mechanical typewriter, and the nearest red telephone box. Today, I can be contacted by my London office anywhere in Scotland by 'bleeper' or mobile phone. I write my copy on a laptop computer, which allows me to communicate with London and the world via a mobile phone. I could, I often muse, send my stories electronically from a beach, rather than an office. (I did once, in some style, from Copacabana in Rio de Janeiro during the Earth Summit in 1992!).

The revolution in the means of communication means that we know about the destruction of the Brazilian rainforest as it happens, and, suddenly, our environmental guilt can be global in scale. But while recent decades have been epitomised by increasing concern, our fears have often been expressed through membership of the Royal Society for the Protection of Birds (RSPB) and the National Trust, rather than lifestyle changes. Our philosophical commitment to a better environment has yet to be converted into action.

In the specifially Scottish context, most of us recognise there are too many people suffering from fuel poverty, that the transport network is dominated by polluting road traffic, and that our fishing stocks are being over-exploited. If the answer to such problems is education, then we will find in years to come that the Peterhead fisherman stops landing black fish, the family escaping from Glasgow for a day does not dump its litter on the banks of Loch Lomond, and the Highland gamekeeper stops destroying the eggs of hen harriers. But it may be another decade, or two, before the environment ethic already found in Scandinavian countries is evident here. If I am wrong, and if the changes required in agriculture, forestry and transport, for example, do not continue apace, then Scotland's special qualities will be much diminished.

The issues to be addressed, practical and philosophical, are legion. It would be good, for example, to see our education system deal with the enmities between town and country; it would be good to see an end to 'townie' protests on grouse moors, led by people who have no knowledge of the issues. The great countryside rally in Hyde Park, in July 1997, revealed ignorance on both sides of the town versus country debate. It was particularly depressing to hear speakers delivering illiberal judgements on their urban neighbours. The right-wing media, which lost its influence in the wake of the Labour general election victory, was beside itself with admiration for those 'from all classes and walks of life' who marched to London to support fox-hunting. One newspaper said the impeccable behaviour of the country folk, who did not drink alcohol and left hardly a scrap of litter, revealed the yawning chasm between those who live in high-rise flats and the gentle souls of good old, rural Britain. That may have been snobbery verging on fascism, but the very fact that such distinctions are drawn between people who live in rural areas and those who live in cities is an indication of the simplistic nature of current environmental debate.

Eventually, all environment issues will become apolitical. I would like to see more people living in the countryside, and I would like to see great chunks of Scotland's wildest lands protected from further development, unless it is the development of the natural habitat. I would also like to see more community ownership, and the development of co-operative, mixed farming/forestry communities.

The realisation of this vision night involve the Cairngorms and Loch Lomond being given national park status – the latter is already assured – and crofting land being owned and managed by crofters.

In the hard landscape of the businessman, education must deliver the message that sound environmental policies can be sound financial policies, but also that the traditional profits-at-all-costs ambition is bankrupt. The concept of sustainable development must pervade business decisions, international relations and trade, agriculture and town planning. Robin Cook's assertion in his early months at Foreign Secretary that this central goal would inform Foreign Office thinking was important.

In a sustainable society, we will understand Edmund Burke's comment, to the effect that 'no man made a bigger mistake than he who did nothing, because he could only do a little'. It has been claimed, for example, that if everyone in Britain swapped one 100-watt lightbulb for a low-energy bulb, it would save the power output and emissions of one medium-sized power station.

In a sustainable world, we will appreciate that waste not put unthinkingly in our refuse bins does make a difference, and that our choices as consumers and shareholders do make a difference. In the new stakeholder society, we will have an opportunity to influence business decisions.

Eventually, I guess taxation will no longer be personal, but will be based on non-renewable resources, while polluting processes are taxed almost out of existence. There is no shortage of precedent: Germany has a car tax based on exhaust fumes and noise, rather than engine size, Finland taxes single-hulled oil tankers because they are more likely to spill their contents; Denmark has increased taxation on raw materials and tripled its tax on rubbish, and Norway introduced a deposit on new cars in 1978, which is returned with a bonus if the car is taken to an approved site at the end of its life.

I guess that the transport systems in our cities will be radically different and that agriculture and aquaculture, which may have expanded dramatically to replace some sea fishing, will be more finely tuned to society's needs. I guess that design standards in houses and public buildings will be of a much higher order, and will be nearly standardised throughout the developed world. I imagine that one country will not be radically greener than the next, as we become more homogeneous in our sense of obligation to the environment.

Simon Pepper, of the Worldwide Fund for Nature, explains the nature of modern life by pointing out that 300 years ago the residents of Aberfeldy, his home town, would have sourced their material requirements within a radius of several miles. Today, the plastic garden toys are from Taiwan, the television is from Japan, the car is from Germany, the vegetables from Italy, and so on. The marketplace will remain global, but perhaps we should find ways to prevent farmed venison from New Zealand being exported to the UK, to be sold for less than Scotland's excellent, and organic, venison. We might decide that more of our needs can be met within narrower boundaries.

Using the concept of 'environmental space', most bulk goods – staple foods, energy, raw materials – might be sourced within the expanded Europe, with specialist items coming from further afield. The EU is currently said to 'occupy' 276 000 square kilometres of agricultural land outside Europe, in addition to its own farmland, to meet its demands. At the same time, some forms of trade will have to be more relaxed than they are today. In a sustainable world, the developed regions must share their new technologies, particularly environmental and energy technologies, with developing nations.

An impractical utopia? We have to hope not. Those who are not convinced by such a positive prognosis, might reflect that modern society has many successes to celebrate, including space exploration, the eradication of many fatal diseases and vastly improved health care.

The alternative to such radical change would mean extracting every last drop of oil and burning it to release its locked-up carbon; letting the North grow richer and the South poorer; continuing to dump waste in the sea and underground; the building of more motorways, and motorways on top of motorways; and global warming swamping Pacific islands.

Home Truths
.

In Scotland, we need to move away from the 'protected area' system in the countryside, to produce policies which will encompass nature all the way from the town centre to the mountain summit. We should not be aiming for islands of biodiversity in a sea of mediocrity.

Agriculture as we understand it today will become less important, but subsidies will drive environmental improvements, including the expansion of hedgerows and small wetlands, the replacement of tree lines and dry stone dykes, and the expansion of semi-natural forests. Farmers will become, in the true sense of the expression, land managers and also custodians. Most of Europe's food will be produced in the areas offering the richest soils and best growing conditions.

Much grazing land will be subject to forest regeneration projects offering the prospect of increased tourism, hunting, craft materials and timber from native species. Some extinct mammals will be reintroduced, and the population at large will support the changes, because it understands them. Scotland's stock among the international environment community will rise.

Commercial woodlands will be increasingly attractive to visitors, and wildlife, and Scotland may resolve eventually to grow Norway spruce – which would have been native but for the loss of the land bridge with Europe – but not the more exotic lodgepole pine, nor even the fast-growing North American sitka spruce. Young trees, in the sustainable landscape, will not be dependent on fencing for their survival. The current deer population will be drastically reduced, while woodlands expand. Consequently, Scotland's red deer will be heavier, healthier animals with grander antlers, and the cost of deer stalking will rise proportionately. Where private landowners fail to achieve their annual cull targets for red deer, the Deer Commission will round up deer in corrals, or send in its own men in helicopters, to complete the cull.

Sheep, there being less of them, will be less vilified, and the descendants of today's shepherds may be tomorrow's farmer/foresters. There may be a shift on the land from food and fibre production, to conservation and amenity goals.

Planning officers will consider environmental and social sustainability hand-in-hand – and before any other consideration – and partnerships will flourish across county boundaries, with the lines between private and public business becoming increasingly blurred. The countryside around towns will be developed, enhanced and safeguarded by such arrangements, which will also help to protect Scotland's important firths.

It goes without saying that Scottish National Heritage (SNH) and the Scottish Environment Protection Agency (SEPA) will be larger, better-funded, more independent bodies. Conservation scientists will know how a given number of birds of prey interact with a given number of red grouse on a sporting estate and will put in place mechanisms to avoid conflict. They will also work towards the restoration of sea trout and salmon stocks in a programme involving international research on migratory fish at sea, and in their North Atlantic breeding grounds.

Large areas of sensitive soil will still be recovering from the effects of acid deposition, but international agreements will be in place and emissions from English power stations will have been cut through cleaning technology and through the growth of renewable energy and energy efficiency measures.

The existing view that Scotland is largely a land of unspoiled mountains, clean air, and pristine lochs, the Scotland of a million brochures, will be much nearer the reality than it was back at the end of the second millennium AD.

We will be able to look back at the 1990s and recognise that decade as a watershed in environmental policy and attitudes; a time when schools and colleges and universities were producing the chief executive officers of the future who would have very different priorities from their grandparents.

Who knows, perhaps the roots of the great Environmental Revolution will be traced to New Year's Day in 1973, when Britain acceded to the Treaty of Rome, or to the Single European Act of 1986, which stated that environmental protection and improvement, the protection of human health and prudent use of natural resources were 'fundamental'. After all, between 1973 and 1997, the EU agreed almost 300 separate measures to protect the environment, including regulations to combat acid rain, traffic pollution, water pollution, the export of hazardous waste, and the decline in species and habitats.

World Truths
.

The world's problems are about population and equity. The issues are people, debt, global warming, ozone depletion, over-fishing, the dumping of waste in oceans, the development of the polar regions, the loss of species and habitats and the use of non-renewable resources. Sustainability means living on the earth's income, not eroding its capital. On a global level, the environmental conundrum comes down to the fact that the industrial nations, with less than one quarter of the world's population, consume 80 per cent of its resources, and generate 75 per cent of the waste.

Currently, many of the countries facing the most serious environmental degradation are also facing serious economic difficulties and debt burdens accrued through the 1980s. The debt puts pressure on developing countries to strip their natural resources – tropical forest in many instances – and uncontrolled population growth exacerbates the situation. Some progress has been made through so-called debt-for-nature swaps, in which debtor countries provide currency for conservation measures, and environmental organisations buy some of the foreign debt from creditors at a discount. We need to reverse the ludicrous situation in which the South gives the North more in debt repayment than it receives in aid. In 1970, most developed nations promised to give 0.7 per cent of their GNP to overseas aid. Today, only a handful, not including the UK or the USA, have achieved that figure.

Some people believe that population alone is the greatest environmental issue. It took the whole of human history up to 1850, when my great-grandfather was alive, for the world population to reach one billion. The first billion, of course, is always the most difficult. Today we are expanding at the rate of one billion per decade. The population doubled between 1950 and 1987, to five billion. Around 94 per cent of the 90 million people added to the world population every year – 250 000 a day – are born in developing countries, where farmers are forced to cultivate marginal land, the soil deteriorates, the demand for wood causes deforestation and drought kills millions. Birth control projects in Africa and Asia, and the education of women in developing countries, are, therefore, central to global environmental strategies.

All over the developing world we can see, on a grand scale, the mistakes the developed world made last century being repeated. The mass migration to towns and cities from the countryside has created appalling slums and desperate river and air pollution. I mentioned my visit to the Earth Summit in Rio, and the joys of Copacabana, but one aspect of modern Brazil had more effect on me than all the talk of the politicians and environmentalists. Thousands of street children, some no more than toddlers, live and die on the streets of Rio every day, having fled the dysfunctional families of the favellas, the crazy do-it-yourself slums made from packing crates and stolen wood and construction waste which climb, like human ant hills, up the steepest slopes on the edge of the city, their water and electricity stolen from mains supplies, their people scratching an existence dominated by prostitution, disease, poverty and violence.

One small boy who said he was twelve, but looked like eight, asked me to be his dad and take him home. He had been sleeping in doorways for four years and survived by begging, stealing and scavenging, and by avoiding the assassins hired by businessmen to shoot children like him and clean up the street. I wrote at the time about a boy with an angelic face who visited a Catholic refuge on a daily basis to have a bowl of soup and to escape the hardness of the street. I read a few years later that he had been shot dead. No-one was charged.

Around the world, there are an estimated 100 million children working as traders, beggars, prostitutes and petty criminals, and 35 million of them live on the street. Every tourist flying, at great expense, to holiday in Rio de Janeiro will meet these children in the open-air cafes and bars, asking for food and drink, and asking, unknowingly, much larger questions. By the year 2000, almost half the world's population is expected to live in cities. In 1990, the figure was only 14 per cent.

My first encounter with the problems of the developing world was in the tiger economy of Thailand, where the steaming, heavily polluted capital of Bangkok, its skyline bristling with cranes and five-star hotels in the making, is thousands of years, but just a two-hour flight, from its Stone Age hinterland.

On the Thai–Burmese border, I spent two days walking in the forest with a native guide and met villagers who had only seen a few white people. They cleared small patches of the forest to cultivate crops, gathered fruit, trapped small animals, carried water from a stream in bamboo pitchers, and lit their cooking fires each day from glowing embers.

One family let me sleep on the rice store outside the door of their hut. In the morning I was woken, while it was still dark, by a cockerel, a girl of five lighting the day's fire from the night's embers, and by the pounding of grain being turned into floor by stone against wood.

The experience was fascinating and poignant. I was a tourist, one of the first, but one of many to come. Now, a few years later, the people of the village may have been strongly influenced by the outside world, and drawn, inexorably, towards its material riches. They may already be selling trinkets and coveting the transistor radio they heard on the forest road.

Isn't it extraordinary that, at the end of this millennium, it is still possible to discover forest tribes which have had no contact with the outside world, existing in the same era as the overwhelmingly technological cities of Tokyo and New York? Early in the third millennium AD, we may lose the last peoples who have a life-or-death understanding of the natural world. While we still can, we must learn from them, and consider our fascination with such a primitive way of life, which leaves no damaging footprint, while we each consume and pollute on a vast scale.

Postscript
·

Between AD 995 and 1000, Scandinavia converted to Christianity just in case the turn of the millennium meant something apocalyptic. In 1001, with the coast clear, Scandinavia relapsed into paganism. This time, with millennial excitement again in the air, our conversion to a less damaging way of life needs to be more profound.

Bibliography

·

Association for the Conservation of Energy, *Energy Efficiency Opportunities in the UK Electricity Sector*, 1992

J. A. Baddeley, D. B. A. Thompson and J. A. Lee, *Regional and Historical Variation in the Nitrogen Content of Racomitrium lanuginosum in Britain in Relation to Atmospheric Nitrogen Deposition*, 1994

D. Begg, 'Innovative Thinking in Transport Planning', speech to 21st Nottingham Transport Conference, 1997

D. Begg, 'Reducing Car Use', speech at Transport and Air Pollution Conference, 1997

D. Begg, 'Transporting Edinburgh into the 21st Century', speech to Light Rail 96 Conference

L. R. Brown *et al.*, *State of the World 1997*, London, 1997

Centre for Environment and Business in Scotland, *Waste Minimisation in Scotland*, Edinburgh, 1997

Centre for Environment and Business in Scotland, *Developing Successful Business with the Environment in Mind*, Edinburgh, 1996

City of Edinburgh Council, Edinburgh's Way Ahead, city development strategy

Conservation Organisations, *Biodiversity Challenge*, 1994

A. Cramb, *Who Owns Scotland Now?*, Edinburgh, 1996

A. Cran and J. Robertson, *Dictionary of Scottish Quotations*, Edinburgh, 1996

T. C. D. Dargie and D. J. Briggs, *State of the Scottish Environment 1991*, Scottish Wildlife and Countryside Link, Perth, 1991

Energy Action Scotland, *Energy Review*, May 1997

Energy Action Scotland, *16 Energy Efficiency Projects*, 1997

Energy Action Scotland, *Fuel Poverty Across Scotland*, 1996

Energy Action Scotland, *Warm Homes are a Basic Right*, 1996

Forestry Authority, *The UK Forestry Standard*, Draft for Consultation, June 1996

Forest Enterprise, *Glen Affric Forest*, 1994

Freshwater Fisheries Laboratory, *Statistical Bulletin*, Fisheries Series, Edinburgh, 1996

Friends of the Earth Scotland, *Hormone Disrupting Compounds*, 1996

Friends of the Earth Scotland, *Towards a Sustainable Scotland*, Edinburgh, 1996

Friends of the Earth Scotland, *Safe Energy Journal*, March–May 1996

Friends of the Earth Scotland, *Achieving the Possible, A Sustainable Energy Strategy*, Edinburgh, 1997

Friends of the Earth Scotland, *Planning Renewables*, Edinburgh, 1997

Friends of the Earth Scotland, *The State of Scotland's Air*, Edinburgh, 1996

Greenpeace, *Industrial Fisheries, from Fish to Fodder*, 1996

Greenpeace Briefing, *Products Based on Industrial Fisheries*, 1996

S. Gubbay, *Scottish Environment Audits 1, The Marine Environment*, Scottish Wildlife and Countryside Link, 1997

T. Hart, 'The Role of Public Transport in Scotland: From Now to 2020', paper to CIT Conference, 1997

HMSO, *The Scottish Environment Statistics*, Edinburgh, 1996

HMSO, *Sustainable Development, The UK Strategy*, London, 1994

HMSO, *Sustainable Forestry, The UK Programme*, 1994

HMSO, *This Common Inheritance, Britain's Environmental Strategy*, London, 1990

Hydro-Electric, *Hydro-Electric Development in Scotland and its Effect on Fish*, Perth, 1996

M. Magnusson, 'The 20th TB Macaulay Lecture', Aberdeen, 1996

P. S. Maitland, P. J. Boon and D. S. McLusky, *The Fresh Waters of Scotland*, Chichester, 1994

A. C. McKinnon, *Logistics, Peripherality and Manufacturing Competitiveness*, 1997

A. C. McKinnon, *Logistical Restructuring, Freight Traffic Growth and the Environment*, 1997

A. C. McKinnon, *The Contribution of Road Construction to Economic Development*, 1996

A. C. McKinnon, 'The empty running and return loading of road goods vehicles', *Transport Logistics*, Vol 1, 1996

National Farmers' Union of Scotland, *Policy Options for Scottish Agriculture*, 1997

Bibliography

RSPB, *A Management Guide to Birds of Scottish Farmland*, 1995

RSPB, *Persecution, A Review of Bird of Prey Persecution in Scotland*, 1996

RSPB, *Wildlife and Agriculture in Scotland: A Secure Future*, 1996

SAC, Perth, *Loch Leven Catchment Management Project*, D. Flint, 1996

Safe Energy Journal, February 1997

Scottish Environmental Forum, *Greening the Grey, Proceedings of a Conference on Habitat II and Sustainable Settlements*, 1996

Scottish Environmental Protection Agency, *Corporate Plan, 1996/97–1999/2000*

Scottish Environment Protection Agency, *Draft National Waste Strategy: Scotland, 1997*

Scottish Environment Protection Agency, *Marine Cage Fish Farming in Scotland*

Scottish Environment Protection Agency, *State of the Environment Report 1996*

Scottish Hydro-Electric, *1996/97 Environmental Report*

Scottish Native Woods, *Why Manage Riparian Woodlands?*, 1996

Scottish Natural Heritage, *Boglands, Scotland's Living Landscapes*, 1995

Scottish Natural Heritage, *The Environment – Who Cares*, Roger Crofts, 1995

Scottish Natural Heritage, *The Natural Heritage of Scotland: An Overview*, Edinburgh, 1995

Scottish Natural Heritage, *Annual Report, 1995–96*

The Scottish Office, *Habitats and Birds Directives, Circular 6*, Edinburgh, 1995

The Scottish Office, *Rural Framework, Scotland's Coasts*, 1996

The Scottish Office, *Statistical Bulletin, Waste Collection, Disposal and Regulation Statistics 1994*

ScottishPower, *Environment Report 1996–97*

Scottish Wildlife and Countryside Link, *National Parks for Scotland*, a paper by Robert Aitken, 1996

Scottish Wildlife Trust, *Head on Collision Scotland*, 1995

J. B. Sissons, *The Evolution of Scotland's Scenery*, Edinburgh, 1966

T. C. Smout, *A Century of the Scottish People 1830–1950*, London, 1987

A. C. Stevenson and D. B. A. Thompson, *Long-term Changes in the Extent of Heather Moorland in Upland Britain and Ireland: Palaeoecological Evidence for the Importance of Grazing*

Strathclyde Regional Council, *Sustainability Indicators*, Glasgow, 1995

D. B. A. Thompson, 'Biodiversity in montane Britain: habitat variation, vegetation diversity and some objectives for conservation', *Biodiversity and Conservation*, vol. 1, 1992

D. B. A. Thompson and D. Horsfield, *Upland Habitat Conservation in Scotland: A Review of Progress and Some Proposals for Action*, 1997

E. O. Wilson, *The Diversity of Life*, London, 1993

S. J. Woodin and A. M. Farmer, *The Effects of Acid Deposition on Nature Conservation in Britain*, Nature Conservancy Council, 1991

WWF Scotland, *Wild Rivers*, a discussion paper by Dr Peter Maitland, 1996

Index

Index